Speech Enhancement, Modeling and Recognition

Speech Enhancement, Modeling and Recognition

Editor

Danel Jaso

scitus
academics

Speech Enhancement, Modeling and Recognition
Edited by **Danel Jaso**

ISBN: 978-1-68117-585-0
Library of Congress Control Number: 2016946440

SCITUS Academics LLC,
616, Corporate Way, Suite 2, 4766,
Valley Cottage, NY 10989

www.scitusacademics.com

Notice

Printed in United States of America on Acid Free Paper

Preface

Communication via speech is one of the essential functions of human beings. Humans possess varied ways to retrieve information from the outside world or to communicate with each other and the three most important sources of information are speech, images and written text. For many purposes, speech stands out as the most efficient and convenient one. Speech not only conveys linguistic contents, but also communicates other useful information like the mood of the speaker. When speaker and listener are near to each other in a quiet environment, communication is generally easy and accurate. However, at a distance or in a noisy background, the listener's ability to understand suffers. Speech enhancement aims to improve speech quality by using various algorithms. The objective of enhancement is improvement in intelligibility and/or overall perceptual quality of degraded speech signal using audio signal processing techniques. Enhancing of speech degraded by noise, or noise reduction, is the most important field of speech enhancement, and used for many applications such as mobile phones, VoIP, teleconferencing systems, speech recognition, and hearing aids.

Speech Enhancement, Modeling and Recognition covers important fields in speech processing such as speech enhancement, noise cancellation, multi resolution spectral analysis, voice conversion, speech recognition and emotion recognition from speech in addition to applications. This book will be of immense useful for advanced graduate students, researchers and practicing engineers employed in speech processing.

Table of Contents

Chapter 1

Esophageal Speech Enhancement Using a Feature Extraction Method Based on Wavelet Transform

Alfredo Victor Mantilla Caeiros[1] and Hector Manuel Pérez Meana[2]

[1] Tecnológico de Monterrey, Campus Ciudad de Mexico, México

[2] Instituto Politécnico Nacional, México

1. INTRODUCTION

People that suffer from diseases such as throat cancer require that their larynx and vocal cords be extracted by surgery, requiring then rehabilitation in order to be able to reintegrate to their individual, social, familiar and work activities. To accomplish this, different methods have been suggested, such as: The esophageal speech, the use of tracheoesophageal prosthetics and the Artificial Larynx Transducer (ALT), also known as "electronic larynx" [1, 2].

The ALT, which has the shape of a handheld device, introduces an excitation in the vocal tract by applying a vibration against the external walls of the neck. The excitation is then modulated by the movement of the oral cavity to produce the speech sound. This transducer is attached to the speaker's neck, and in some cases to the speaker's cheeks. The ALT is widely recommended by voice rehabilitation physicians because it is very easy to use even for new patients, although the voice produced by these transducers is unnatural and with low quality, besides that it is distorted by the ALT produced background noise. Thus, ALT results in a considerably degradation of the quality and intelligibility of speech, problem for which an optimal solution has not yet been found [2].

The esophageal speech, on the other hand, is produced by the compression of the contained air in the vocal tract, from the stomach to the mouth through the esophagus. This air is swallowed and it produces a vibration of the esophageal upper muscle as it passes through the esophageal-larynx segment, producing the speech. The generated sound is similar to a burp, the tone is commonly very low, and the timbre is generally harsh. As in the ALT produced speech, the voiced segments of esophageal speech are the most affected parts of the speech within a word or phrase resulting an unnatural speech. Thus many efforts have been carried out to improve its quality and intelligibility.

Several approaches have been proposed to improve the quality and intelligibility of alaryngeal speech, esophageal as well as ALT produced speech [2, 3].

This chapter presents an alaryngeal speech enhancement system, which uses several methods for speech recognition such as voiced and unvoiced segment detection, feature extraction method and pattern recognition algorithms.

The content of this chapter is a follows:

1) Acquisition and preprocessing of esophageal speech: This section explains the acquisition and preprocessing of speech signal including filtering, segmentation and windowing.

2) Voiced/unvoiced segment detection: This section discusses several methods for classifying voiced and unvoiced segments such as:

    - Pitch detection

    - Zero crossing

    - Formant Analysi

3) Feature extraction: The performance of any speech recognition algorithm strongly depends on the accuracy of the feature extraction method. This section exposes some important feature extraction methods such as the Linear Predictive Coding (LPC), the Cepstral Coefficients, as well as a feature extraction method based on an inner

ear model, which takes into account the fundamental concepts of critical bands using a wavelet function. The later method emulates the basilar membrane operation, through a multiresolution analysis similar to that performed by a wavelet transform.

4) Clasiffier: The parameter vector obtained in the feature extraction stage is supplied to a classifier. The classification stage consists of neural networks, which identifies the voiced segments present in segment under analysis.

5) Voice synthesis: The voiced segments detected are replaced by voiced segments of a normal speaker and concatenated with unvoiced and silent segments to produce the restored speech.

6) Results: Finally, using objective and subjective evaluation methods, it shows that the proposed system provides a fairly good improvement of the quality and intelligibility of alaryngeal speech signals.

2. METHODS

Figure 1 shows a block diagram of the proposed system. It is based on the replacement of voiced segments of alaryngeal speech by their equivalent normal speech voiced segments, while keeping the unvoiced and silence segments without change. The main reason is that the voiced segments have a more significant impact on the speech quality and intelligibility than the unvoiced segments.

The following explains the stages of the system.

2.1. Data Acquisition

The first stage consists of recording a speech file, from an esophageal speaker. The audio data is kept as a WAV file. All files are monaural and they are digitalized in PCM (Pulse Code Modulation) format using a sampling rate of 8000Hz with a resolution of 8bits.

2.2. Preprocessing

The digital signal is low-pass filtered to reduce the background noise. This stage is implemented by a 200 order digital FIR filter with a cut-off frequency of 900Hz. A common practice for speech recognition is the use of pre-emphasis filter in order to amplify the higher frequency components of the signal with the purpose of emulating the additional sensibility of the human ear to high frequencies. Generally a high pass filter characterized by a slope of 20 dB per decade is used [4].

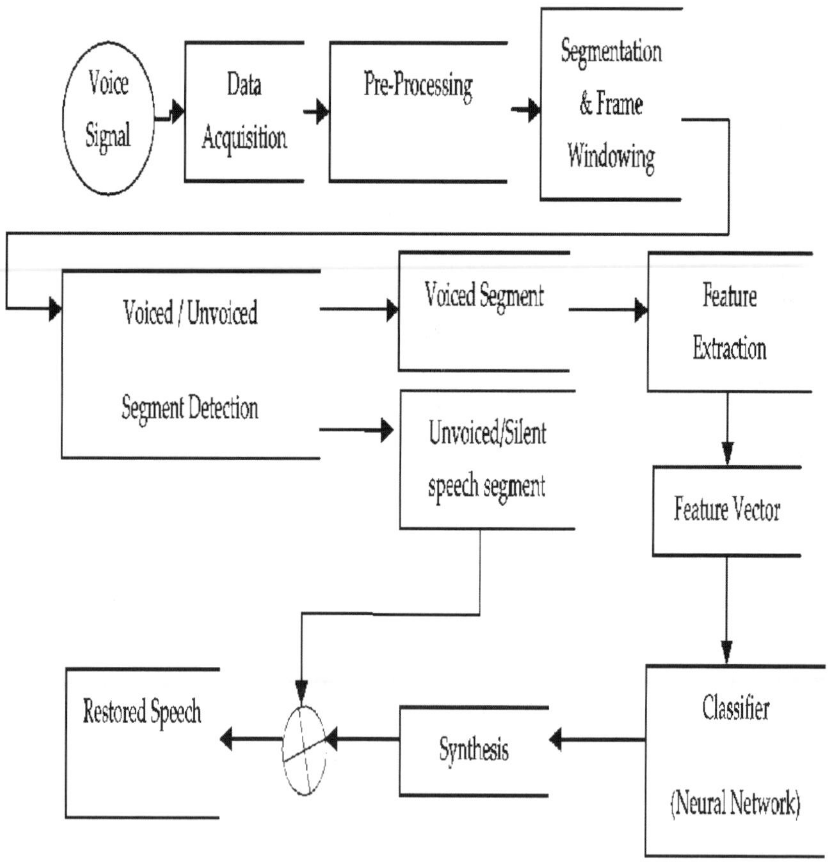

Figure 1. Block Diagram

2.3. Segmentation and Frame Windowing

The filtered signal is divided into 100ms segments (800 samples) and at the same time each segment is subdivided into 10ms blocks, on which the later processing is going to be realized. The size of these blocks is determined by the quasi-periodic and quasi-stationary character of speech in such interval.

A Hamming window is applied to the segmented signal so that the extreme samples of the segments had less weight that the central samples. The window's length is chosen to be larger than the frame interval, preventing a loss of information which could take place during the transitions from one frame to the next.

2.4. Voiced/Unvoiced Segment Detection

A voiced (sonorous) segment is characterized by a periodic or quasiperiodic behavior in time, a fine harmonic frequency structure produced by the vibration of the vocal chords, as well as a high energy concentration due to the little obstruction that the air meets in its way through the vocal tract. The vowels and some consonants present such behavior.

Several approaches have been proposed to detect the voiced segments of speech signals. However the use of a single criterion of decision to determine if a speech segment is voiced or unvoiced is not enough. Thus most algorithms in the speech processing area use the combination of more than one criterion. The proposed speech restoration method uses the combination of energy average, zero crossing and formant analysis of speech signal for voiced/unvoiced segment classification

2.4.1. Energy Average

A first criterion ponders the average power of each frame by comparing it to that of its surroundings. An interval of 100 milliseconds in the neighborhood of the actual frame was selected. As part of the system's initial configuration, two thresholds are fixed. If the quotient between the frame's average power and that of the frame's surrounding is smaller than the lower threshold, the frame is labeled as unvoiced. Otherwise, if the quotient is larger than the higher threshold, the frame would be taken as voiced. For those cases in which the rate of average power lies between both thresholds, the energy criterion is not enough to determine the signal's nature.

2.4.2. Zero Crossing

The second criterion is based on the signal periodicity using the number of zero crossings in each frame. Here two thresholds are used to establish that in a noise free speech segment of 10ms a voiced segment has about 12 zero crossings, while in an unvoiced segment has about 50 zero crossings [5, 6]. These values are not fixed and must be adjusted according to the sampling frequency used. In the proposed algorithm, for a sampling frequency of 8 kHz the maximum value of zero crossings that could be detected in 10ms is approximately 40. Thus an upper threshold of 30 was chosen for voiced/unvoiced classification.

2.4.3. Formant Analysis

The third criterion is based on the amplitude of formants which, represents the resonance frequency of the vocal tract. Formants are the envelope peaks of the speech signal power spectrum density. The frequencies in which the first formants are produced are of great importance in speech recognition [4].

The formants are obtained from the polynomial roots generated by the linear prediction coefficients (LPC) that represent the vocal tract filter. Once the formants, whose frequency is defined by the angle of the roots closer to the unitary circle, are obtained, they are ordered in an ascending form and the first three formants are chosen as parameters of the speech segment. These formants are then stored in the system so that they can be employed to take the voiced/invoiced decision. Using the normalized Fast Fourier Transform (FFT) the amplitude of the formant frequency can be obtained.

To take the decision whether the segment is voiced or not, the value of the formants amplitude is normalized each 100 millisecond segment. Then the algorithm finds the maximum value of each formant among the 10 values stored for each fragment. Then each value is divided between the estimated maximum values as shown in (1).

$$AF_1 = \left[\frac{AF_{1-1}}{AF_{1Max}} \frac{AF_{1-2}}{AF_{1Max}} \cdots\cdots \frac{AF_{1-10}}{AF_{1Max}} \right]$$

$$AF_2 = \left[\frac{AF_{2-1}}{AF_{2Max}} \frac{AF_{2-2}}{AF_{2Max}} \cdots \frac{AF_{2-10}}{AF_{2Max}} \right]$$

$$AF_3 = \left[\frac{AF_{3-1}}{AF_{3Max}} \frac{AF_{3-2}}{AF_{3Max}} \cdots\cdots \frac{AF_{3-10}}{AF_{3Max}} \right] \quad (1)$$

The local normalization process is justified for esophageal speakers due to the loss of energy as they speak. Once the normalized values are obtained, the decision is made using an experimental threshold value which is equal to 0.25. It can be seen as a logic mask in the algorithm if the normalized values greater than 0.25 are set to one, otherwise are set to zero, as shown in (2).

$$\frac{AFx - N}{AFx \max} = \begin{cases} 0 & \dfrac{AFx - N}{AFx \max} < 0.25 \\ 1 & \dfrac{AFx - N}{AFx \max} > 0.25 \end{cases} \quad (2)$$

formant array using the values obtained after the threshold operation. Here only the segments in which the three formants have values over the 0.25 are considered to be voiced segments.

Finally, using the three criterions mentioned above, a window is applied to the original signal which is equal to one if the segment is classified as voiced by the three methods; and it is equal to zero otherwise, such that only the voiced segments of the original signal are obtained.

2.5. Feature Vector Extraction

The performance of any speech recognition algorithm strongly depends on the accuracy of the feature extraction method. This fact has motivated the development of several efficient algorithms to estimate a set of parameters that allow a robust characterization of the speech signal. Some of these methods are: The Linear Prediction Coefficients (LPCs), Formants Frequencies Analysis, Mel Frequency Cepstral Coefficients (MFCC) among others [5,6]. This section discusses these methods and proposes one based on Wavelet Transform.

2.5.1. Linear Prediction Coefficients (Lpcs)

The LPCs methods are based on the fact that the signal can be approximated from a weighted sum of precedent samples [7]. This approximation is given by:

$$s'_n = \sum_{k=1}^{p} a_k s_{n-k} \tag{3}$$

where a_k (1<k<p) is a set of real constants known as predictor coefficients, that must be calculated, and p is the predictor order. The problem of linear prediction resides on finding the predictor coefficients a_k that minimize the error between the real value of the function and the approximated function.

To minimize the total quadratic error is necessary to calculate the autocorrelation coefficients. This is a matrix equation with different recursive solutions, the commonly used is the Levinson recursion.

The developed algorithm takes each segment of 10 milliseconds and calculates its linear prediction coefficients. The number of predictor coefficients is obtained by substituting the sampling frequency value (*fs*) in (4).

$$p = 4 + \frac{f_s}{1000} = 4 + \frac{8000}{1000} = 12 \tag{4}$$

The sequence of the minimal error could be interpreted as the output of the H(z) filter when it is excited by the S_n signal. H(z) is usually known as an inverted filter. The approximated transfer function could be obtained if it is assumed that the transfer function S(z) of the signal is modeled as an only pole filter with the form of (5).

$$\hat{S}(z) = \frac{A}{H(z)} = \frac{A}{1 - \sum_{k=1}^{p} a_k z^{-k}} \quad (5)$$

The LPC coefficients correspond to Sˆ(z)S^(z) poles. Therefore, the LPC analysis aims to calculate the filter properties of the vocal tract that produces the sonorous signal.

2.5.2. Formant Frequencies Analysis

Formants are the envelope peaks of the speech signal spectrum that represent the resonance frequency of the vocal tract.

If the spectrum of a speech signal can be approximated only by its poles, then the formants could be obtained from the Sˆ(z)S^(z) poles. The poles of Sˆ(z)S^(z) can be calculated making the denominator of (5) to zero and solving it to find its roots. The S plane conversion is done by substituting z byeskTeskT, where $s_k$is the pole in the s plane. The resultant roots are generally conjugated complex pairs.

Formants frequencies are obtained from the polynomial roots generated by the linear prediction coefficients. The formant frequency is defined by the angle of the roots closer to the unitary circle. A root, with an angle close to zero, indicates the existence of a formant near the origin. A root whose angle is in close proximity to π indicates that the formant is located near the maximum frequency, in this case 4000Hz. Since the frequency dominion is symmetric with respect to the vertical axis, the roots located in the inferior semi plane of z plane can be ignored.

Let r_p as a linear prediction coefficient root with real part φp and imaginary part

$$r_p = \phi_p + j\theta_p \quad (6)$$

The roots (r) which are located in the superior semi plane near the unitary circle can be obtained using (7).

$$r_p = \begin{cases} 0 & for \quad \theta_p \leq 0.01 \\ r_p & for \quad \theta_p > 0.01 \end{cases} \tag{7}$$

By using the arctangent function is possible to obtain the roots angle. Doing this, the roots are mapped into the frequency dominion by using (8) to get the formants.

$$for_p = \vartheta_p \frac{f_s}{2\pi} \tag{8}$$

Once the formants are obtained, they are organized in ascending order, and the first three are chosen as parameters of the speech segment.

2.5.3. Mel Frequency Cepstral Coefficients (Mfcc)

The cepstral coefficients estimation is another widely used feature extraction method in speech recognition problems. These coefficients form a very good features vector for the development of speech recognition algorithms, sometimes better than the LPC ones.

Cepstrum is defined as the inverse Fourier transform of spectrum module logarithm (9)

$$c(t) = F^{-1}\left[\log\left(S(\omega)\right)\right] \tag{9}$$

Developing de above equation it obtains:

$$c(t) = F^{-1}\left\{\left[\log E(\omega)\right] + \left[\log H(\omega)\right]\right\} \tag{10}$$

The above equation indicates that Cepstrum of a signal is the sum of Cepstrum excitation source and the vocal tract filter. The vocal tract information is of slow variation, and it appears in the first cepstrum coefficients. For speech recognition application the vocal tract information is more important than excitation source. The cepstral coefficients can be estimated from the LPC coefficients applying the following recursion:

$$c_o = \ln \sigma^2$$
$$c_m = a_m + \sum_{k=1}^{m-1} \left(\frac{k}{m} \right) c_k a_{m-k} \qquad 1 \le m \le p$$
$$c_m \sum_{k=1}^{m-1} \left(\frac{k}{m} \right) c_k a_{m-k} \qquad m > p \qquad (11)$$

where *Cm* is the n-th LPC-Cepstral coefficients, *a* is the i-th LPC coefficients and m is the Cepstral index.

Usually the number of cepstral coefficients is equal to the number of LPC ones to avoid noise. A representation derived from the coefficients cepstrum are the Mel Frequency Cepstral Coefficients (MFCC) whose fundamental difference with Cepstrum coefficients is that the frequency bands are positioned according to a logarithmic scale known as MEL scale, which approximates the frequency response of the human auditory system more efficiently than Fast Fourier Transform(FFT).

2.5.4. Feature Extraction Method Based On Wavelet Transform

Most widely used feature extraction methods, such as those described above, are based on modeling the form in which the speech signal is produced. However if the speech signals are processed taking into account the form in which they are perceived by the human ear, similar or even better results may be obtained. Thus using an ear model-based feature extraction method might represent an attractive alternative, since this approach allows characterizing the speech signal in the form that it is perceived [8]. This section proposes a feature extraction method based on an inner ear model, which takes into account the fundamentals concepts of critical bands.

In the inner ear, the basilar membrane carries out a time-frequency decomposition of the audible signal through a multiresolution analysis similar to that performed by a wavelet transform. Thus to develop a feature extraction method that emulates the basilar membrane operation, it should be able to carry out a similar frequency decomposition, as proposed in the inner ear model developed by Zhang et. al [9]. In this model the dynamics of basilar membrane, which has a characteristic frequency equal to *fc*, can be modeled by using a gamma-tone filter which consists of a gamma distribution multiplied by a pure tone of frequency *fc*. The shape of the gamma distribution α, is related to the filter order while the scale ϑ, is related to period of occurrence of the events under analysis, when they have a Poisson distribution. Thus the gamma-tone filter representing the impulse response of the basilar membrane is given by (12)

$$\psi_{\theta}^{\alpha}(t)=\frac{1}{(\alpha-1)\,!\,\theta^{\alpha}}t^{\alpha-1}e^{\frac{-t}{\theta}}\cos(2\pi t/\theta) \quad t>0 \tag{12}$$

Equation (12) defines a family of gamma-tone filters characterized by and . Thus to emulate the basilar membrane behavior, it is necessary to look for the more suitable filter bank which, according to the basilar membrane model given by Zhang [9], can be obtained if we set $\vartheta=1$ and $\alpha=3$, *since* such values (12) result in the best approximation to the inner ear dynamics. From (12) we have:

$$\psi(t)=\frac{1}{2}t^{2}e^{-t}\cos(2\pi t) \quad t>0 \tag{13}$$

Taking the Fourier transform of (13)

$$\Psi(\omega)=-\frac{(\omega-2\pi)^{2}}{\left[1+j(\omega-2\pi)\right]^{3}}+\frac{(\omega+2\pi)^{2}}{\left[1+j(\omega+2\pi)\right]^{3}} \tag{14}$$

It can be shown that *ψ (t)* presents the expected attributes of a mother wavelet since it satisfies the admissibility condition given by (15)

$$\int_{-\infty}^{\infty}\frac{\left|\Psi(\omega)\right|^{2}}{|\omega|}d\omega<\infty \tag{15}$$

This means that *ψ (t)* can be used to analyze and then reconstruct a signal without loss of information [10]. That is the functions given by (13) constitute an unconditional basis in $L^2(\mathbf{R})$ and then we can estimate the expansion coefficients of an audio signal *f(t)* by using the scalar product between *f(t)* and the function *ψ(t)* with translation τ and scaling factor *s* as follows:

$$C(\tau,s)=\frac{1}{\sqrt{s}}\int_{0}^{\infty} f(t)\psi\left(\frac{t-\tau}{s}\right)dt \tag{16}$$

A sampled version of (16) must be specified because we require characterizing discrete time speech signals. To this end, a sampling of the scale parameter, s, involving the psychoacoustical phenomenon known as critical bandwidths will be used [11].

The critical bands theory models the basilar membrane operation as a filter bank in which the bandwidth of each filter increases as its central frequency also increases. This requirement can be satisfied using the Bark frequency scale that is a logarithmic scale in which the frequency resolution of any section of the basilar membrane is exactly equal to one Bark, regardless of its characteristic frequency. Because the Bark scale is characterized by a biological parameter, there is not an exact expression for it given as a result several different proposals available in the literature. Among them, the statistical fitting provided by Schroeder et al [11], appears to be a suitable choice. Thus using the approach provided, the relation between the linear frequency, *f* in Hz and the Bark frequency *Z*, is given by

$$Z=7\ln\left(\frac{f}{650}+\sqrt{\left(\frac{f}{650}\right)^{2}+1}\right) \tag{17}$$

Using (17) the *j-th* scaling factor s_j given by the inverse of the *j-th* central frequency in Hz, f_c,corresponding to the *j-th* band in the Bark frequency scale becomes

$$s_j=\frac{e^{j/7}}{325\left(e^{2/7}-1\right)},\quad j=1,2,3,.... \tag{18}$$

The inclusion of bark frequency in the scaling factor estimation, as well as the relation between (17) and the dynamics of basilar membrane, allows frequency decomposition similar to that carried out by the human ear. Since the scaling factor given by (18) satisfies the Littlewood-Paley theorem (19)

$$\lim_{j\to+\infty}\frac{s_{j+1}}{s_j}=\lim_{j\to+\infty}\frac{e^{(j+1)/7}\left(e^{2j/7}-1\right)}{e^{j/7}\left(e^{2(j+1)/7}-1\right)}=e^{-1/7}\neq 1$$
(19)

there is not information loss during the sampling process. Finally the number of subbands is related to the sampling frequency as follows

$$j_{\max}=\operatorname{int}\left(7\ln\left(\frac{f_s}{1300}+\sqrt{\left(\frac{f_s}{1300}\right)^2+1}\right)\right)$$
(20

Therefore, for a sampling frequency equal to 8 KHz the number of subbands becomes 17. Finally, the translation axis is naturally sampled because the input data is a discrete time signal and then the *j-th*decomposition signal can be estimated as follows

$$C_j(m)=\sum_{-\infty}^{\infty}f(n)\psi_j(n-m)$$
(21)

where

$$\psi_j(n)=\frac{1}{2}\left(\frac{nT}{s_j}\right)^2 e^{-\left(\frac{nT}{s_j}\right)}\cos\left(\frac{2\pi nT}{s_j}\right)\quad n>0$$
(22)

In (22) T denotes the sampling period. The expansion coefficients *Cj* obtained for each subband are used to estimate the feature vector to be used during the training and recognition tasks.

Using (21), the feature vector used for voiced segment identification consists of the following parameters:

a) The energy of the *m-th,* speech signal frame $\overline{x^2(n)}$, where $1\le n\le N$, and N is number of samples in the *m-th* frame.

b) The energy contained in each one of the 17 wavelet decomposition levels of *m-th* speech frame C2j⁻⁻⁻⁻⁻(m) Cj2⁻(m) , where

1≤j≤17

c) The difference between the energy of the previous and actual frames given by (23)

$$d_x(m) = \overline{x^2}(n - mN) - \overline{x^2}(n - (m-1)N) \quad (23)$$

d) The difference between the energy contained in each one of the 17 wavelet decomposition levels of current and previous frames given by (24)

$$\overline{v}_j = \overline{c_j^2}(m) - \overline{c_j^2}(m-1) \quad (24)$$

The last eighteen members of the feature vector include the spectral dynamics of speech signal concatenating the variation from the past feature vector to the current one.

$$\mathbf{X}(m) = \left[\overline{x^2}(n - mN), \overline{c_1^2}(m), \overline{c_2^2}(m), \ldots, \overline{c_{17}^2}(m), d_x(m), \overline{v}_1(m), \overline{v}_2(m), \ldots, \overline{v}_{17}(m)\right] \quad (25)$$

2.6. Classification Stage

The classification stage consists of one neural network, which identifies the vowel, in cascade with a parallel array of 5 neural networks, which are used to identify the alaryngeal speech segment to be changed by its equivalent normal speech segment, as shown in Figure 2. At this point, the estimated feature vector, given by (25), is feed into the first ANN (Figure 2) to estimate vowel present in the segment under analysis. Once the vowel is identified, the same feature vector is feed into the five ANN structures of the second stage, along with the output of first ANN, to identify the vowel-consonant combination contained in the voiced segment under analysis. The output of enabled ANN corresponds to the codebook index of identified segment. Thus the first ANN output is used to enable the ANN corresponding to the detected vowel, disabling the other four while the second ANN is used to identify the vowel-consonant or vowel-vowel combination. The ANN in the first stage has 10 hidden neurons while the ANNs in the second stage have 25.

The ANN training process is carried out in two steps. First the ANN used to identify the vowel contained in the speech segment is trained in a supervised manner using the backpropagation algorithm. After convergence is achieved, the enabled ANN in the second stage is used to identify the vowel-consonant or vowel-vowel combination and is also trained in a supervised manner using the backpropagation algorithm, while the coefficients vectors of the other 4 ANN are kept constant. In all cases 650 different alaryngeal voiced segments with a convergence factor equal to 0.009 are used, achieving a global mean square error of 0.1 after 400,000 iterations.

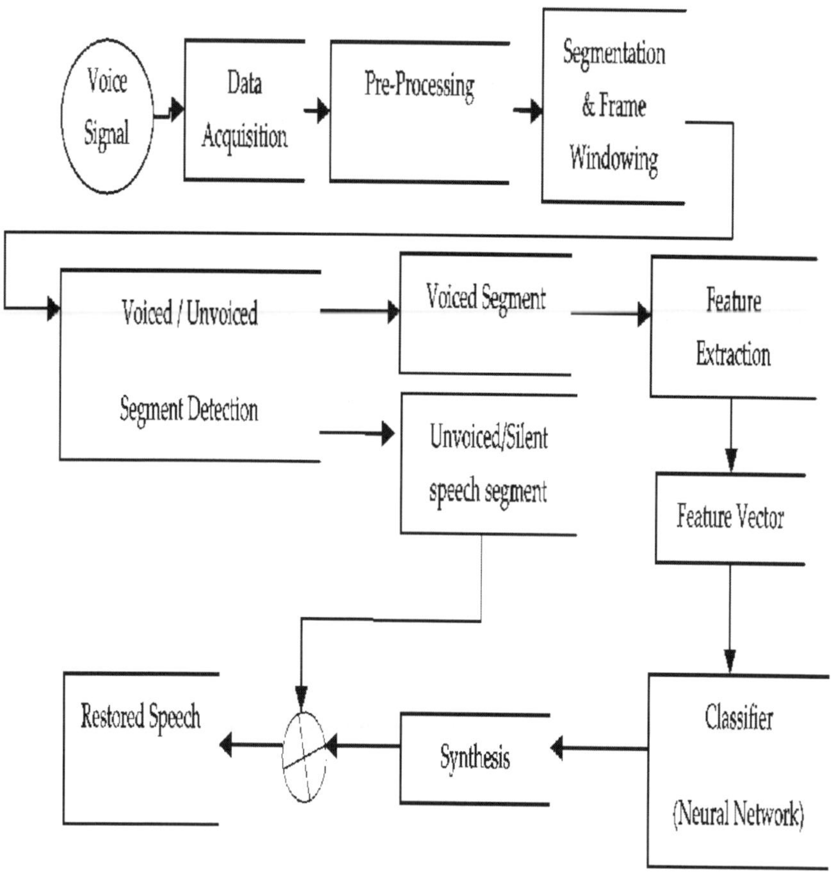

Figure 2. Pattern recognition stage. The first ANN indentifies the vowel present in the segment and the other 5 ANN identify the consonant-vowel combination.

2.7. Synthesis Stage

This stage provides the restored speech signal. According to Figure 1, if a silence or unvoiced segment is detected, the switch is enabled and the segment is concatenated with the previous one to produce the output signal. If voice activity is detected, the speech segment is analyzed using the energy analysis, the zero crossings number and the formant analysis explained in section 2.4. If a voiced segment is detected, it is identified using pattern recognition techniques (ANN). Then the alaryngeal voiced segment is replaced by the equivalent normal speech voiced segment, contained in the codebook, which is finally concatenated with the previous segments to synthesize the restored speech signal.

3. RESULTS

Figure 3 shows the plot of mono-aural recordings of Spanish words "abeja" (a), "adicto" (b) and "cupo" (c), pronounced by an esophageal speaker with a sample frequency of 8 kHz, respectively, including the detected voiced segments. Figure 3 shows that a correct detection is achieved using the combination of several features, in this case zero crossing, formants analysis and energy average.

Figure 4 shows the produced esophageal speech signal corresponding to the Spanish word "cachucha" (cap) together with the restored signal obtained using the proposed system. The corresponding spectrograms of both signals are shown in Figure 5.

To evaluate the actual performance of the proposed system, two different criteria were used: the bark spectral distortion (MBSD) and the mean opinion scoring (MOS). The bark spectrum L(f) reflects the ear's nonlinear transformation of frequency and amplitude, together with the important aspects of its frequency and spectral integration properties in response to complex sounds. Using the Bark spectrum, an objective measure of the distortion can be defined using the overall distortion as the mean Euclidian distance between the spectral vectors of the normal speech, $L_n(k,i)$, and the processed ones, $L_p(k,i)$, taken over successive frames as follows.

$$MBSD = \frac{\sum_{k=1}^{N}\sum_{i=1}^{M}\left[L_n(k,i) - L_p(k,i)\right]^2}{\sum_{k=1}^{N}\sum_{i=1}^{M}L_n^2(k,i)} \quad (26)$$

where L*n*(k,i) is the Bark spectrum of the kth segment of the original signal, L*p*(k,i) is the Bark spectrum of the processed signal and M is the number of critical bands. Figures 6 and 7 show the Bark spectral trace of both, the esophageal speech produced and enhanced signals, respectively corresponding to the Spanish words "hola" (hello) and "mochila" (bag). Here the MBSD during voiced segments was equal 0.2954 and 0.4213 for "hola" and "mochila", respectively, while during unvoiced segments the MBSD was 0.6815 and 0.7829 for "hola" and "mochila" respectively. The distortion decreases during the voiced periods as suggested by (26). Evaluation results using the Bark spectral distortion measures show that a good enhancement can be achieved using the proposed method.

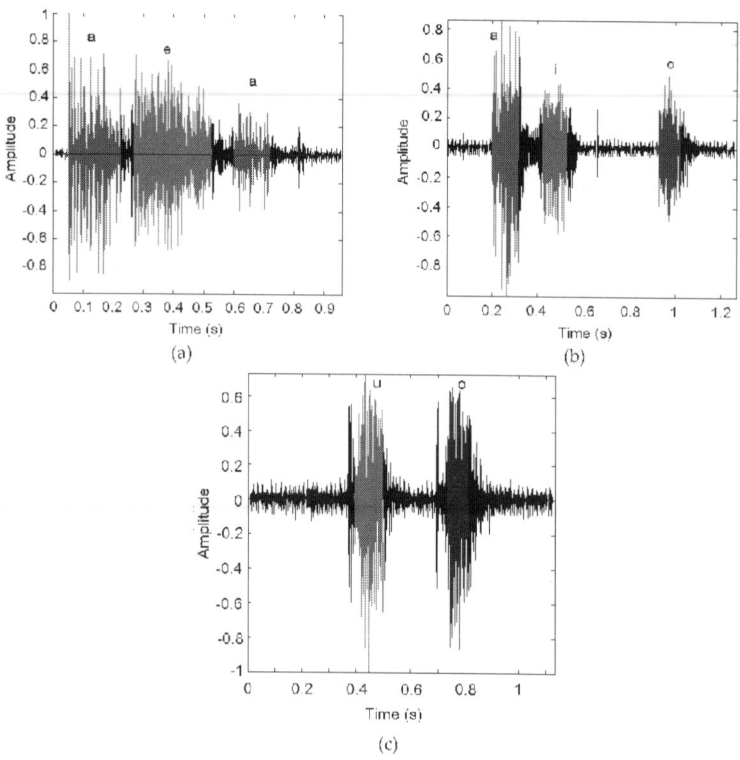

Figure 3. Detected voiced/unvoiced segments of esophageal speech signal of Spanish words "abeja" (a), "adicto" (b) and "cupo" (c).

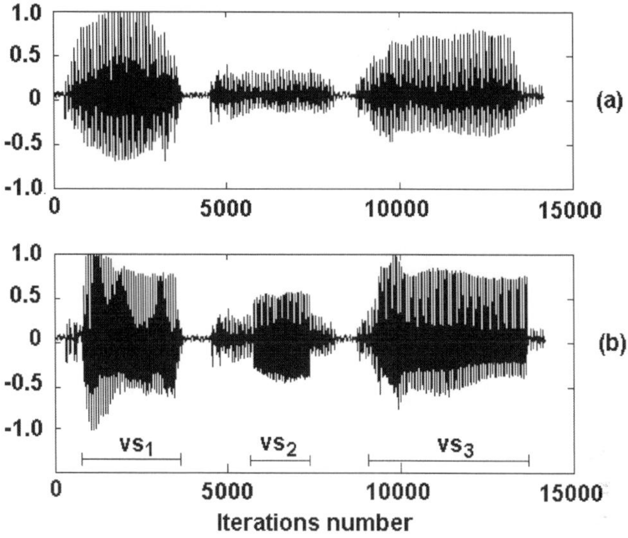

Figure 4. Waveforms trace corresponding to the Spanish word, "Cachucha", (Cap). a) produced Esophageal speech, b) restored speech.

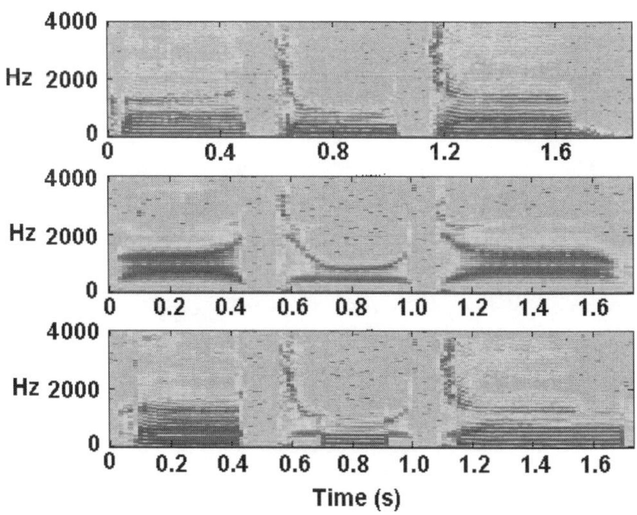

Figure 5. Spectrograms trace corresponding to the Spanish word, "Cachucha" (Cap). a) Normal speech, b) Produced Esophageal Speech, c) Restored speech.

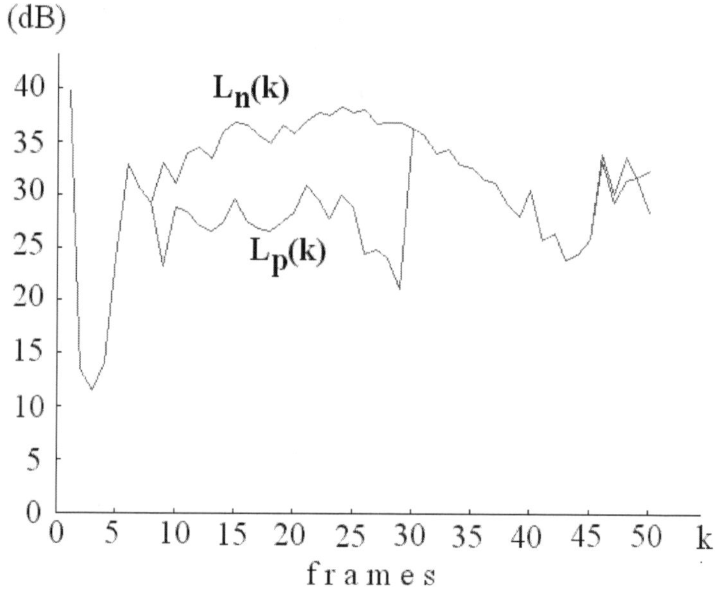

Figure 6. Bark spectral trace of normal, $L_n(k)$, and enhanced, $L_p(k)$, speech signals of the Spanish word "hola".

A subjective evaluation was also performed using the Mean Opinion Scoring (MOS) in which the proposed system was evaluated by 200 normal speaking persons and 200 alaryngeal ones (Table 1 andTable 2), from the point of view of intelligibility and speech quality where 5 is the highest score and 1 is the lowest one. In both cases the speech intelligibility and quality evaluation without enhancement are shown for comparison. These evaluation results show that the proposed system improves the performance of [2] which reports a MOS of 2.91 when the enhancement system is used and 2.3 without enhancement. These results also show that, although the improvement is perceived by the alaryngeal and normal speakers, the improvement is larger in the opinion of alaryngeal speakers. Thus the proposed system is expected to have a quite good acceptance among the alaryngeal speakers, because the proposed system allows synthesizing several kinds of male and female speech signals.

Finally, about 95% of alaryngeal persons participating in the subjective evaluation preferred the use of the proposed system during conversation.

Subjective evaluation shows that quite a good performance enhancement can be obtained using the proposed system.

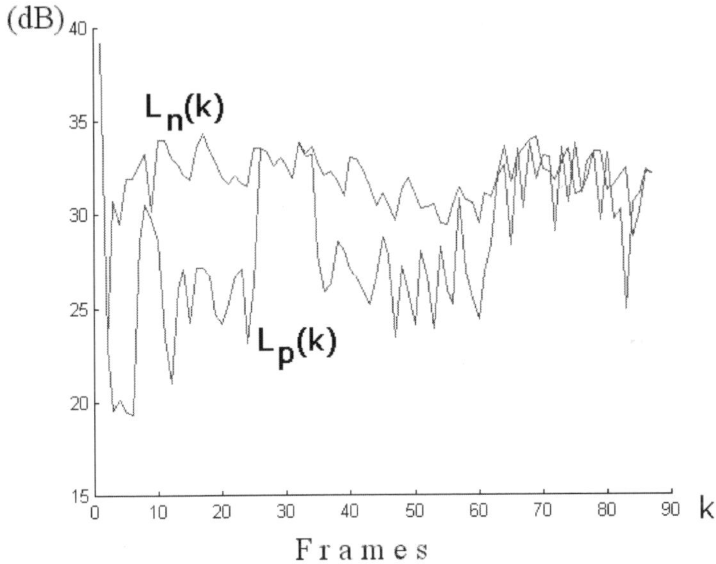

Figure 7. Bark spectral trace of normal, $L_n(k)$, and enhanced, $L_p(k)$, speech signals of the Spanish word "mochila".

Table 1. Subjective evaluation of esophageal speech without enhancement.

	Normal listener		Alaryngeal listener	
	Quality	Intelligibility	Quality	Intelligibility
MOS	2.30	2.61	2.46	2.80
Var	0.086	0.12	0.085	0.11

The performance of the voiced segments classification stage was evaluated using 450 different alaryngeal voiced segments. The system failed to classify correctly 22 segments, which represents a misclassification rate of about 5% using a network as identification method, while a misclassification of about 7% was obtained using the Hidden Markov Models (HMM). The comparison results are given in Table 3.

Table 2. Subjective evaluation of proposed alaryngeal speech enhancement system

	Normal listener		Alaryngeal listener	
	Quality	Intelligibility	Quality	Intelligibility
MOS	2.91	2.74	3.42	3.01
Var	0.17	0.102	0.16	0.103

Table 3. Recognition performance using two different identification methods using the feature extraction method based on wavelet transform.

Identification Method	Normal Speech	Alaryngeal Speech
ANN	98%	95%
HMM	97%	93%

The behavior of proposed feature extraction method was compared with the performance of several other wavelet functions for evaluation purposes. Comparison results are shown in Table 4 which show that proposed method has better performance than other wavelet based feature extraction methods.

Table 4. Performance of different wavelet based feature enhanced methods when an ANN is used as identification method.

	Proposed method	Daub 4 wavelet	Haar wavelet	Mexican hat wavelet	Morlet wavelet
Recognition rate	95%	75%	40%	79%	89%

4. CONCLUSIONS

This chapter proposed an alaryngeal speech restoration system, suitable for esophageal and ALT produced speech, based on a pattern recognition approach where the voiced segments are replaced by equivalent segments of normal speech contained in a codebook. Evaluation results show a correct detection of voiced segment by comparison between their spectrograms to those spectrograms of normal speech signal. Objective and subjective evaluation results show that the proposed system provides a good improvement in the intelligibility and quality of esophageal produced speech signals. These results show that proposed system is an attractive alternative to enhance the alaryngeal speech signals. This chapter also presents a flexible structure that allows the use of the proposed system to enhance esophageal and artificial laryinx produced speech signals without further modifications. The proposed system could be used to enhance alaryngeal speech in several practical situations such as telephone and teleconference systems, thus improving the voice and quality life of alaryngeal people.

REFERENCES

1. H. K. Barney, H. L. Hawork, F. E. , Dunn, 1959 1959), "An experimental transitorized artifcial larynx",. Bell System Technical Journal, 38 13371356 .

2. G. Aguilar, M. Nakano-Miyatake, H. Perez-Meana, 2005 Alaryngeal Speech Enhancement Using Pattern Recognition Techniques", IEICE Trans. Inf. & Syst. E88-D , 7 16181622 .

3. D. Cole, S. Sridharan, M. Geva, 1997 "Application of noise reduction techniques for alaryngeal speech enhancement", IEEE TECON, Speech and Image Processing for Computing and Telecommunications, 491494.

4. H. David, et.al, 2001 "Acoustics and psychoacoustics", Ed: Focal Press. Second Edition.

5. L. Rabiner, B. Juang, 1993 "Fundamentals of Speech Recognition", Prentice Hall, Piscataway, USA.

6. L. R. Rabiner, B. H. Juang, C. H. Lee, 1996 An Overview of Automatic Speech Recognition", in Automatic Speech and Speaker Recognition: Advanced Topics, C. H. Lee, F. K. Soong and K. K. Paliwal editors, Kluwer Academic Publisher, 130 , Norwell MA.

7. D. G. Childers, 2000 "Speech Processing and syntesis toolboxes", Wiley & Sons, inc.

8. A. Mantilla-Caeiros, M. Nakano-Miyatake, H. Perez-Meana, 2007 "A New Wavelet Function for Audio and Speech Processing", 50th MWSCAS, 101104 .

9. X. Zhang, M. Heinz, I. Bruce, L. Carney, 2001 "A phenomenological model for the responses of auditory-nerve fibers: I. Nonlinear tuning with compression and suppression", Acoustical Society of America, 109 2 648670 .

10. R. M. Rao, A. S. Bopardikar (1998),"Wavelets Transforms, Introduction to Theory and Applications", Addison Wesley, New York.

11. M. R. Schroeder, 1979 "Objective measure of certain speech signal degradations based on masking properties of the human auditory perception", Frontiers of Speech Communication Research, Academic Press, New York.

Chapter 2

Robust Distributed Speech Recognition Using Auditory Modelling

Ronan Flynn[1] and Edward Jones[2]

[1] School of Engineering, Athlone Institute of Technology, Athlone, Ireland

[2] College of Engineering and Informatics, National University of Ireland, Galway, Ireland

1. INTRODUCTION

The use of the Internet for accessing information has expanded dramatically over the past few years, while the availability and use of mobile hand-held devices for communication and Internet access has greatly increased in parallel. Industry has reacted to this trend for information access by developing services and applications that can be accessed by users on the move. These trends have highlighted a need for alternatives to the traditional methods of user data input, such as keypad entry, which is difficult on small form-factor mobile devices. One alternative is to make use of automatic speech recognition (ASR) systems that act on speech input from the user. An ASR system has two main elements. The first element is a front-end processor that extracts parameters, or features, that represent the speech signal. These features are processed by a back-end classifier, which makes the decision as to what has been spoken.

In a fully embedded ASR system [1], the feature extraction and the speech classification are carried out on the mobile device. However, due to the computational complexity of high-performance speech recognition systems, such an embedded architecture can be impractical on mobile hand-held terminals due to limitations in processing and memory resources. On the other hand, fully centralised (server-based) ASR systems have fewer computational constraints, can be used to share the computational burden between mobile users, and can also allow for the easy upgrade of speech recognition technologies and services that are provided. However, in a centralised ASR system the recognition accuracy can be compromised as a result of the speech signal being distorted by low bit-rate encoding at the codec and a poor quality transmission channel [2, 3].

A distributed speech recognition (DSR) system is designed to overcome some of the difficulties described above. In DSR, the terminal (the mobile device) includes a local front-end processor that extracts, directly from the speech, the features to be sent to the remote server (back-end) where recognition is performed. In mobile environments, the speech features can be sent over an error protected data channel rather than a voice channel, making the DSR system more robust to channel errors.

However, DSR systems generally operate in high levels of background noise (particularly in mobile environments). For mobile users in noisy environments (airports, cars, restaurants etc.) the speech recognition accuracy can be reduced dramatically as a consequence of additive background noise. A second source of error in DSR systems is the presence of transmission errors in the form of random packet loss and packet burst loss during transmission of speech features to the classifier. Packet loss can arise in wireless and packet switched (IP) networks, both networks over which a DSR system would normally be expected to operate. Packet loss, in particular packet burst loss, can have a serious impact on recognition performance and needs to be considered in the design of a DSR system.

This chapter addresses the issue of robustness in DSR systems, with particular reference to the problems of background noise and packet loss, which are significant bottlenecks in the commercialisation of speech recognition products, particularly in mobile environments. The layout of the chapter is as follows. Section 2 discusses the DSR architecture and standards in more detail. This is followed by an overview of the auditory model used in this chapter as an alternative front-end to those published in the DSR standards. The Aurora 2 database and a description of the speech recognition system used are also discussed in Section 2. Section 3 addresses the problem of robustness of speech recognition systems in the presence of additive noise, in particular, by examining in detail the use of speech enhancement techniques to reduce the effects of noise on the speech signal. The performance of a DSR system in the presence of both additive background noise and packet loss is examined in Section 4. The feature vectors produced by the auditory model are transmitted over a channel that is subject to packet burst loss and packet loss mitigation to compensate for missing features is investigated. Conclusions are presented in Section 5.

2. DISTRIBUTED SPEECH RECOGNITION SYSTEMS

2.1. Dsr Architecture and Standards

A DSR system is designed as a compromise between local and centralised recognition, in order to alleviate the issues associated with these approaches [2, 3]. In DSR, the speech recognition task is split between the terminal or client, where the front-end feature extraction is performed, and the network or server, where the back-end recognition is performed. The features that represent the speech are sent by means of an error protected data channel to the classifier for processing. DSR avoids both the speech encoding and decoding stages associated with centralised recognition and so eliminates the degradations that originate from the speech compression algorithms. The bandwidth required to transmit the extracted features to the server is much less than what is required to send the encoded speech signal. DSR systems offer some advantages over other architectures. Recent comparative studies have shown the superior performance of DSR to codec-based ASR [4]. However, in a DSR system, transmission errors in the form of random packet loss and packet burst loss still need to be taken into consideration. Such transmission errors can have a significant impact on recognition accuracy.

Furthermore, it is well known that the presence of noise severely degrades the performance of speech recognition systems, and much research has been devoted to the development of techniques to alleviate this effect; this is particularly important in the context of DSR where mobile clients are typically used in high-noise environments (though the same problem also exists for local embedded, or centralised architectures in noisy conditions). One approach that can be used to improve the robustness of ASR systems is to enhance the speech signal itself before feature extraction. Speech enhancement can be particularly useful in cases where a significant mismatch exists between training and testing conditions, such as where a recognition system is trained with clean speech and then used in noisy conditions. A significant amount of research has been carried out on speech enhancement, and a number of approaches have been well documented in the literature [5]. There has also been much interest in DSR in recent years, within the research community, and in international standardisation bodies, in particular, the European Telecommunications Standards Institute (ETSI) [6-9], which has developed a number of different recommendations for front-end processors of different levels of complexity.

The ETSI basic front-end [6] was developed for implementation over circuit-switched channels and this implementation is also considered in the other three standards. The advanced front-end [7] produces superior performance to the basic front-end, and was designed to increase robustness in background noise. The implementation of the front-ends over packet-switched Internet Protocol (IP) networks has been specified in two documents published by the Internet Engineering Task Force (IETF). The first of these [10] specifies the real-time transport protocol (RTP) payload format for the basic front-end while the second [11] specifies the RTP payload format for the advanced front-end.

The ETSI basic and advanced front-ends both implement MFCC-based parameterisation of the speech signal. The stages involved in feature extraction based on MFCCs are shown in Figure 1. The speech signal first undergoes pre-emphasis in order to compensate for the unequal sensitivity of human hearing across frequency. Following pre-emphasis, a short-term power spectrum is obtained by applying a fast Fourier transform (FFT) to a frame of Hamming windowed speech. Critical band analysis is carried out using a bank of overlapping, triangular shaped, bandpass filters, whose centre frequencies are equally spaced on the mel scale. The FFT magnitude coefficients are grouped into the appropriate critical bands and then weighted by the triangular filters. The energies in each band are summed, creating a filter bank vector of spectral energies on the mel scale. The size of this vector of spectral energies is equal to the number of triangular filters used. A non-linearity in the form of a logarithm is applied to the energy vector. The final step is the application of a discrete cosine transform (DCT) to generate the MFCCs.

In the ETSI DSR front-ends, speech, sampled at 8 kHz, is blocked into frames of 200 samples with an overlap of 60%. A logarithmic frame energy measure is calculated for each frame before any processing takes place. In the case of the basic front-end, pre-emphasis is carried out using a filter coefficient equal to 0.97 while the advanced front-end uses a value of 0.9. A Hamming window is used in both the ETSI basic and advanced front-ends prior to taking an FFT. In the ETSI advanced front-end a power spectrum estimate is calculated before performing the filter bank integration. This results in higher noise robustness when compared with using a magnitude spectrum estimate as used in the ETSI basic front-end [12]. The two front-ends both generate a feature vector consisting of 14 coefficients made up of the frame log-energy measure (determined prior to pre-emphasis) and cepstral coefficients C_0 to C_{12}. In order to introduce robustness against channel variations, the ETSI advanced front-end carries out post-processing in the cepstral domain on coefficients C_1 to C_{12} in the form of blind equalisation [13].

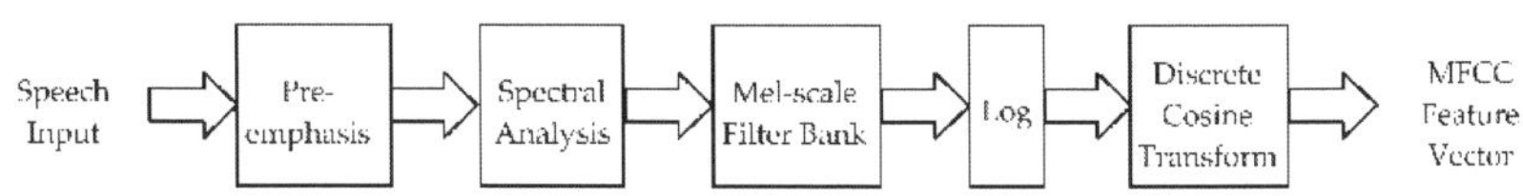

Figure 1. MFCC feature extraction

2.2. Auditory Modelling As An Alternative Front-End

Many computational auditory models have been proposed for use in speech recognition systems, often with excellent results, particularly in the presence of noise. In the work presented here, the auditory model of Li *et al.* [14] is used. The choice of this auditory front-end is motivated by previous work carried out by the authors [15] where a number of auditory front-ends were investigated in a comparative study of robust speech recognition with the widely-used Aurora 2 database [16]. In that study, there was no pre-processing or enhancement of the speech utterances. The front-ends investigated were perceptual linear prediction (PLP) proposed by Hermansky [17], the PEMO algorithm proposed by Tchorz and Kollmeier [18], and the front-end processor proposed by Li *et al.*[14]. For the task of connected digit recognition using the Aurora 2 database, the front-end proposed by Li *et al.* gave the best overall recognition results of all the auditory models examined, and with an overall reduction in recognition error compared to the ETSI basic front-end [6] which was used as a baseline for comparison.

The steps involved in feature extraction in the Li *et al.* auditory model are shown in Figure 2. Speech is sampled at 8 kHz and blocked into frames of 240 samples. Frame overlap is 66.7% and a Hamming window is used prior to taking a FFT. An outer/middle ear transfer function that models pressure gain in the outer and middle ears is applied to the spectrum magnitude. After conversion of the spectrum to the Bark scale, the transfer function output is processed by an auditory filter that is derived from psychophysical measurements of the frequency response of the cochlea. A non-linearity in the form of a logarithm followed by a DCT is applied to the filter outputs to generate the cepstral coefficients. The recognition experiments use vectors that include energy and 12 cepstral coefficients (C_1 to C_{12}) along with velocity and acceleration coefficients. This results in vectors with an overall dimension equal to 39.

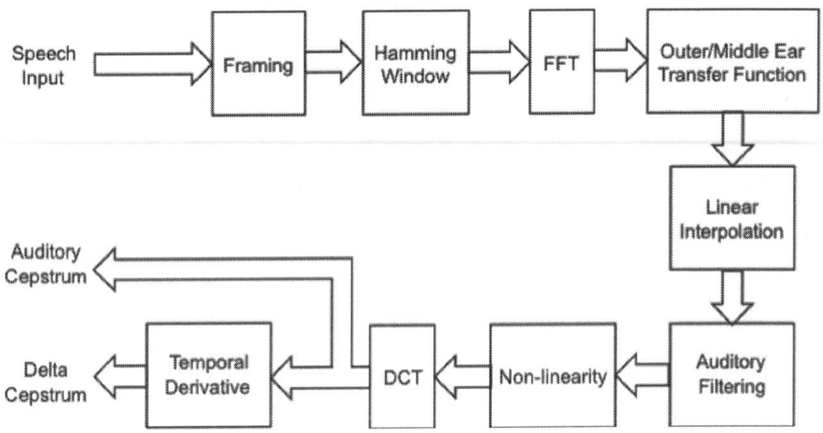

Figure 2. Feature extraction proposed by Li *et al.* [14].

2.3. Aurora 2 Database

The recognition problem examined in this work is connected digit recognition using the Aurora 2 database [16]. The motivation behind the creation of the Aurora database was to provide a framework that allowed for the evaluation and comparison of speech recognition algorithms in noisy conditions, thus providing a good basis for comparison between researchers. It has been widely used in the development and evaluation of DSR systems. The speech database is derived from utterances of isolated digits and connected digit sequences spoken by US-American adults originally included in the well-known TIDigits database. The speech in the TIDigits database is sampled at 20 kHz and is down-sampled to 8 kHz in the Aurora database. Some additional filtering is applied to the down-sampled data in order to take into account the frequency characteristics of equipment used in telecommunications systems. The channel characteristics used are G.712 and modified intermediate reference system (MIRS). The down-sampled, filtered speech corresponds to "clean" data in the Aurora database. The Aurora database also contains "noisy" data. This corresponds to clean data with noise artificially added at SNRs of 20 dB, 15 dB, 10 dB, 5 dB, 0 dB and –5 dB. The noise signals added are chosen to reflect environments in which telecommunication terminals are used. In total there are eight different noise types: subway, babble, car, exhibition hall, restaurant, street, airport and train station.

The Aurora framework includes a set of standard test conditions for evaluation of front-end processors. For the purpose of training the speech recogniser, two modes are defined. The first mode is training on clean data and the second mode is multi-condition training on noisy data. The same 8440 utterances, taken from the training part of the TIDigits, are used for both modes. For the multi-condition training, the clean speech signals are used, as well as speech with four different noise types (subway, babble, car and exhibition hall), added at SNRs of 20 dB, 15 dB, 10 dB and 5 dB. There are three different test sets defined for recognition testing, with the test utterances taken from the testing part of the TIDigits database. Test Set A (28028 utterances) employs the same four noises as used for the multi-condition training. Test Set B uses the same utterances as Test Set A but uses four different noise types (restaurant, street, airport and train station). In both Test Sets A and B, the frequency characteristic used in the filtering of the speech and noise is the same as that used in the training sets, namely G.712. The frequency characteristic of the filter used in Test Set C (14014 utterances) is the MIRS, and is different from that used in the training sets. Subway and street noises are used in Test Set C.

2.4. Speech Recognition System

The classifier used for the recognition experiments in the work presented in this chapter is the HMM-based recogniser architecture specified for use with the Aurora 2 database [16], and implemented with the widely-used HTK package [19]. The use of a well-known specification provides a common framework with which to compare different front-ends and feature vectors for the purpose of connected digit recognition. There are eleven whole word HMMs each with 16 states; each state has 3 Gaussian mixtures. The topology of the models is left-to-right without any skips over states. This topology is suitable for modelling the sequential nature of speech and the consecutive states represent the consecutive speech states in a particular utterance. Two pause models, "sil" and "sp", are defined. The "sil" model has 3 states and each state has 6 mixtures. The "sp" model has a single state. The Baum-Welch re-estimation algorithm is applied in the training of the word models. An utterance can be modelled by any sequence of digits with the possibility of a "sil" model at both ends and adjacent digits separated by a "sp" model.

The method used to measure the performance of a speech recognition system is dependent on the type of utterance that is to be recognised, i.e. isolated word or continuous speech. There are three error types associated with the recogniser in a continuous speech recognition system:

Substitutions (S) – A word in the original sentence is recognised as a different word.

Deletions (D) – A word in the original sentence is missed.

Insertions (I) – A new word is inserted between two words of the original sentence.

The performance measure used throughout the work presented here, and also used in [16], is the word accuracy as defined by (1):

$$Word\ accuracy = \frac{N-(S+D+I)}{N} \times 100\% \qquad (1)$$

where N is the total number of evaluated words. The word accuracies for each of the Aurora test sets presented throughout this chapter are calculated according to [16], which defines the performance measure for a test set as the word accuracy averaged over all noises and over all SNRs between 0 dB and 20dB. The overall word accuracy for the two training modes, clean training and multi-condition training, is calculated as the average over the three test sets A, B and C.

3. SPEECH ENHANCEMENT

Additive noise from interfering noise sources, and convolutional noise arising from transmission channel characteristics both contribute to a degradation of performance in automatic speech recognition systems. This section addresses the problem of robustness of speech recognition systems in the first of these conditions, namely additive noise. As noted previously, speech enhancement is one way in which the effects of noise on the speech signal can be reduced. Enhancement of noisy speech signals is normally used to improve the perception of the speech by human listeners however, it may also have benefits in enhancing robustness in ASR systems. Speech enhancement can be particularly useful in cases where a significant mismatch exists between training and testing conditions, such as where a recognition system is trained with clean speech and then used in noisy conditions, as inclusion of speech enhancement can help to reduce the mismatch. This approach to improving robustness is considered in this section.

In the speech recognition system described here, the input speech is pre-processed using an algorithm for speech enhancement. A number of different methods for the enhancement of speech, combined with the auditory front-end of Li *et al.* [14], are evaluated for the purpose of robust connected digit recognition. The ETSI basic [6] and advanced [7] front-ends proposed for distributed speech recognition are used as a baseline for comparison.

3.1. Speech Enhancement Overview

The enhancement of noisy speech can be described as an estimation problem in which the original clean signal is estimated from a degraded version of the signal. A significant amount of research has been carried out on speech enhancement, and a number of approaches have been well documented in the literature. A survey of a number of approaches to speech enhancement using a single microphone is presented in [5].

Two measures that can be used to perceptually evaluate speech are its *quality* and its *intelligibility* [5]. Speech quality is a subjective measure and is dependent on the individual preferences of listeners. It is a measure of how comfortable a listener is when listening to the speech under evaluation. The intelligibility of the speech can be regarded as an objective measure, and is calculated based on the number or percentage of words that can be correctly recognised by listeners. The intelligibility and the quality of speech are not correlated [5] and it is well known that improving one of the measures can have a detrimental effect on the other one. Speech enhancement algorithms give a trade-off between noise reduction and signal distortion. A reduction in noise can lead to an improvement in the subjective quality of the speech but a decrease in the measured speech intelligibility [5].

When using speech enhancement in an ASR system, the speech is enhanced before feature extraction and recognition processing. The advantage of this is that there is no impact on the computational complexity of the feature extraction or the recognition processes as the enhancement is independent of both, and the speech enhancement can be implemented as an add-on without significantly affecting existing parts of the system. However, every speech enhancement process will introduce some form of signal distortion and it is important that the impact of this distortion on the recognition process is minimised.

3.2. Speech Enhancement Algorithms

In this section, the various speech enhancement algorithms that were examined are briefly described. The algorithms range from well-established algorithms like that of Ephraim and Malah [20], to more recently proposed ones like that of Rangachari and Loizou [21]. Furthermore, the algorithms cover a range of paradigms, including spectral subtraction-based algorithms using the FFT for spectral analysis, as well as methods based on auditory filter banks.

Ephraim and Malah [20] present a minimum mean-square error short-time spectral amplitude (MMSE STSA) estimator. The estimator is based on modelling speech and noise spectral components as statistically independent Gaussian random variables. The enhanced speech is constructed using the MMSE STSA estimator combined with the original phase of the noisy signal. Analysis is carried out in the frequency domain and the signal spectrum is estimated using an FFT.

Westerlund *et al.* [22] present a speech enhancement technique in which the input signal is first divided into a number of sub-bands. The signal in each sub-band is individually multiplied by a gain factor in the time domain based on an estimate of the short term SNR in each sub-band at every time instant. High SNR values indicate the presence of speech and the sub-band signal is amplified. Low SNR values indicate the presence of noise only and the sub-band signal remains unchanged.

Martin [23] presented an algorithm for the enhancement of noisy speech signals by means of spectral subtraction, in particular through a method for estimation of the noise power on a sub-band basis. Martin's noise estimation method is based firstly on the independence of speech and noise, and secondly on the observation that speech energy in an utterance falls to a value close to or equal to zero for brief periods. Such periods of low speech energy occur between words or syllables in an utterance and during speech pauses. The energy of the signal during these periods reflects the noise power level. Martin's minimum statistics noise estimation method tracks the short-term power spectral density estimate of the noisy speech signal in each frequency bin separately. The minimum power within a defined window is used to estimate the noise floor level. The minimum tracking method requires a bias compensation since the minimum power spectral density of the noisy signal is smaller than the average value. In [24], Martin further developed the noise estimation algorithm by using a time- and frequency-dependent smoothing parameter when calculating the smoothed power spectral density. A method to calculate an appropriate time and frequency dependent bias compensation is also described in [24] as part of the algorithm.

Rangachari and Loizou [21] proposed an algorithm for the estimation of noise in highly non-stationary environments. The noisy speech power spectrum is averaged using time and frequency dependent smoothing factors. This new averaged value is then used to update the noise estimate. Signal-presence probability in individual frequency bins is calculated in order to update the smoothing factors. Signal presence is determined by computing the ratio of the noisy speech power spectrum to its local minimum, which is updated continuously by averaging past values of the noisy speech power spectra with a look-ahead factor. The results in [21] indicate that the local minimum estimation algorithm adapts very quickly to highly non-stationary noise environments.

A technique for the removal of noise from degraded speech using two filtering stages was proposed by Agarwal and Cheng [25]. The first filtering stage coarsely reduces the noise and whitens any residual noise while the second stage attempts to remove the residual noise. Filtering is based on the Wiener filter concept and filter optimisation is carried out in the mel-frequency domain. The algorithm, described as a two-stage mel-warped Wiener filter noise reduction scheme, is a major component of the ETSI advanced front-end standard for DSR [7]. The implementation of noise reduction in the ETSI advanced front-end is summarised in [12].

3.3. Tests and Results

This section presents recognition results from tests on the Aurora 2 database [16], using the combination of the speech enhancement algorithms described previously and the auditory model proposed by Li *et al.* (see Section 2.2). In the analysis, two versions of the Li *et al.* front-end are used. The first, referred to as Li *et al.* (I), generates a feature vector consisting of 13 coefficients made up of the frame log-energy measure and the cepstral coefficients C_1 to C_{12}. The second version, referred to as Li *et al.* (II), generates a feature vector that contains the cepstral coefficients C_1 to C_{12} along with a weighted combination of cepstral coefficient C_0 and the frame log-energy measure. The reason for investigating two versions of the Li *et al.* front-end, Li *et al.* (I) and Li *et al.* (II), is to allow for a closer comparison with the ETSI basic front-end [6] and the ETSI advanced front-end [7] respectively. In all cases training was carried out using clean data, so that the effect of the speech enhancement in removing mismatch could be examined. The speech enhancement algorithms were used on both the (clean) training speech as well as the (noisy) test speech. Feature vectors were extracted directly from the enhanced speech with no intermediate processing. The recognition experiments used vectors that include 13 static coefficients along with velocity and acceleration coefficients. This results in vectors with an overall dimension equal to 39. The word accuracies detailed in the tables of results were calculated as previously described in Section 2.4.

In the comparison of Li *et al.* (I) and the ETSI basic front-end, there was no post-processing of the feature vectors carried out. The recognition results using the Aurora 2 database for Li *et al.* (I), for each speech enhancement algorithm, are given in Table 1 and the corresponding results for the ETSI basic front-end (the baseline for this test) are given in Table 2.

The performance of Li *et al.* (II) was compared with the performance of the ETSI advanced front-end. The ETSI advanced front-end includes a SNR-dependent waveform processing block that is applied after noise reduction and before feature extraction. The purpose of this block is to improve the noise robustness in the front-end of an ASR system by enhancing the high SNR period portion and attenuating the low SNR period portion in the waveform time domain, thus increasing the overall SNR of noisy speech [26]. However, the evaluation here is looking primarily at the effect of speech enhancement or noise reduction alone on the connected digit recognition accuracy. Therefore, the waveform processing block in the ETSI advanced front-end was disabled. In addition, the ETSI advanced front-end carries out post-processing in the cepstral domain in the form of blind equalisation as described in [13]. To ensure a closer match with the ETSI advanced front-end, the feature vectors produced by Li *et al.* (II) undergo post-processing in the cepstral domain by means of cepstral mean subtraction (CMS). The recognition results for Li *et al.* (II), for each speech enhancement algorithm, are detailed in Table 3 and the recognition results for the ETSI advanced front-end are detailed inTable 4.

Table 1. Recognition results for the Li *et al.* (I) front-end.

	Absolute word accuracy %			
Enhancement	Set A	Set B	Set C	Overall
None	62.16	64.31	57.76	62.14
Ephraim & Malah	78.85	79.38	74.78	78.25
Westerlund *et al.*	75.87	76.32	70.45	74.97
Martin	72.47	71.96	70.21	71.81
Rangachari & Loizou	74.50	73.16	74.29	73.92
Agarwal & Cheng	86.33	84.87	81.86	84.85

Table 2. Recognition results for the ETSI basic front-end.

	Absolute word accuracy %			
Enhancement	Set A	Set B	Set C	Overall
None	61.34	55.75	66.14	60.06
Ephraim & Malah	76.34	75.91	73.71	75.64
Westerlund *et al.*	76.04	72.54	72.36	73.90
Martin	67.98	67.57	68.24	67.87
Rangachari & Loizou	63.58	61.57	67.82	63.62
Agarwal & Cheng	84.39	82.75	78.72	82.60

Table 3. Recognition results for the Li *et al.* (II) front-end.

	Absolute word accuracy %			
Enhancement	Set A	Set B	Set C	Overall
None	67.34	69.18	63.44	67.30
Ephraim & Malah	80.36	81.03	79.34	80.42
Westerlund *et al.*	78.70	80.02	78.44	79.18
Martin	73.07	72.93	72.17	72.83
Rangachari & Loizou	76.08	76.16	75.94	76.08
Agarwal & Cheng	87.03	86.85	84.58	86.47

Table 4. Recognition results for the ETSI advanced front-end.

	Absolute word accuracy %			
Enhancement	Set A	Set B	Set C	Overall
None	65.92	65.48	70.07	66.57
Ephraim & Malah	77.92	77.61	78.64	77.94
Westerlund *et al.*	79.09	79.13	79.70	79.23
Martin	71.26	72.91	72.71	72.21
Rangachari & Loizou	73.77	73.35	78.85	74.62
Agarwal & Cheng	85.92	85.66	83.89	85.41

Table 5 provides an overall view of the relative performance of the different speech enhancement algorithms for each of the four front-end versions considered.

3.4. Discussion

Ignoring speech enhancement, and comparing Tables 1 and 2, the performance of Li *et al.* (I) exceeds the baseline ETSI front-end [6] by 2.08% overall. From Table 3 and Table 4, again without speech enhancement applied, there is a difference in recognition accuracy of 0.73% in favour of Li *et al.* (II) when compared with the ETSI advanced front-end [7].

Table 5. Performance ranking of enhancement algorithms.

Rank	Li *et al.* (I) FE	ETSI basic FE	Li *et al.* (II) FE	ETSI advanced FE
1	Agarwal & Cheng	Agarwal & Cheng	Agarwal & Cheng	Agarwal & Cheng
2	Ephraim & Malah	Ephraim & Malah	Ephraim & Malah	Westerlund *et al.*
3	Westerlund *et al.*	Westerlund *et al.*	Westerlund *et al.*	Ephraim & Malah
4	Rangachari & Loizou	Martin	Rangachari & Loizou	Rangachari & Loizou
5	Martin	Rangachari & Loizou	Martin	Martin

The other results in Tables 1 to 4 show that enhancement of the speech prior to feature extraction significantly improves the overall recognition performance. This improvement in recognition accuracy is observed for both the ETSI basic [6] and advanced [7] front-ends and the front-end proposed by Li*et al.* [14]. A comparison of Table 1 with Table 2 shows that Li *et al.* (I) outperforms the ETSI basic front-end for all of the speech enhancement techniques evaluated. Furthermore, from Tables 3 and 4, it is seen that Li *et al.* (II) again outperforms the ETSI advanced front-end for all speech enhancement methods except Westerlund *et al.* [22], for which the overall recognition results are quite close.

For Li *et al.* (I), Li *et al.* (II), the ETSI basic front-end and the ETSI advanced front-end, the best overall recognition accuracy is obtained for speech enhancement using the algorithm proposed by Agarwal and Cheng [25]. The combination of auditory front-end and the two-stage, mel-warped, Wiener filter noise reduction scheme results in an overall recognition accuracy that is approximately 6% better overall compared with the next ranked front-end and speech enhancement combination. After Agarwal and Cheng [25], the next best performance across the board is obtained using Ephraim and Malah [20], and Westerlund *et al.* [22]. This suggests that the choice of speech enhancement algorithm for best speech recognition performance is somewhat independent of the choice of front-end (though this would have to be validated by further testing with other front ends).

4. ROBUSTNESS TO NOISE AND PACKET LOSS

In the previous section, the benefit of speech enhancement prior to feature extraction in a speech recognition system was demonstrated. However, in a DSR system, transmission errors can still have a significant impact on recognition performance. Such transmission errors in the form of bit errors, random packet loss and packet burst loss need to be taken into consideration. This is particularly important in the context of increasing use of packet-based networks for transmission of speech and data in mobile environments. This section examines the performance of a DSR system in the presence of both background noise and packet loss.

4.1. Channel Models and Loss Compensation

A DSR client and server may be interconnected over either a circuit-switched channel or a packet-switched channel. Approaches used in the literature to simulate different channel types fall into two categories. The first makes use of physical layer models that simulate transmission phenomena that occur on the physical channel. The second category involves the use of statistical models that model unconditional packet loss probability and conditional packet loss burst lengths. This is the approach used in this chapter.

To simulate packet loss and error bursts, the 2-state Gilbert model is widely used. In [27-29], a voice over IP (VoIP) channel is simulated using such a model. References [30, 31] simulate IP channels and use a 2-state Gilbert model to simulate burst type packet loss on the channel. Statistical models have also been used to simulate the physical properties of the communication channel. The Gilbert model was found in [3] to be inadequate for simulating a GSM channel and instead a two-fold stochastic model is used in which there are two processes, namely shadowing and Rayleigh fading. This same model was used by [32], again to model a GSM network. Reference [33] compares three models of packet loss and examines their effectiveness at simulating different packet loss conditions. The models are a 2-state Markov chain, the Gilbert-Elliot model and a 3-state Markov chain. The 2-state Markov chain in [33] uses State 1 to model a correctly received packet and State 2 to model a lost packet. While the Gilbert-Elliot model is itself a 2-state Markov model, there is only a probability of packet loss when in State 2. The three models in [33] are all validated for GSM and wireless local area network (WLAN) channels.

Results indicate that the 3-state Markov model gives the best results overall and this model is used in the work described here; the model is described in more detail later in this chapter.

There are a number of techniques documented that are used within DSR systems for the purpose of reducing transmission error degradation and so increasing the robustness of the speech recognition. Error-robustness techniques are categorised in [4] under the headings client-based error recovery, and server-based error concealment. Client-based techniques include retransmission, interleaving and forward error correction (FEC). While retransmission and FEC may result in recovering a large amount of transmission errors, they have the disadvantage of requiring additional bandwidth and introducing additional delays and computational overhead. Server-based methods include feature reconstruction, by means of repetition or interpolation, and error correction in the ASR-decoding stage. Reference [4] provides a survey of robustness issues related to network degradations and presents a number of analyses and experiments with a focus on transmission error robustness.

The work described in [34-38] is focused on burst-like packet loss and how to improve speech recognition in the context of DSR. The importance of reducing the average burst length of lost feature vectors rather than reducing the overall packet loss rate is central to the work in these papers. By minimising the average burst length, the estimation of lost feature vectors is more effective. Reference [34] compared three different interleaving mechanisms (block, convolutional and decorrelated) and found that increasing the degree of interleaving increases the speech recognition performance but that this comes with the cost of a higher delay. It is further suggested in [38] that, for a DSR application, it is more beneficial to trade delay for accuracy rather than trading bit-rate for accuracy as in forward error correction schemes. Reference [35] combines block interleaving to reduce burst lengths on the client side with packet loss compensation at the server side. Two compensation mechanisms are examined: feature reconstruction by means of nearest neighbour repetition, interpolation and maximum a-posteriori (MAP) estimation; and a decoder-based strategy using missing feature theory. The results suggest that for packet loss compensation, the decoder-based strategy is best. This is especially true in the presence of large bursts of losses as the accuracy of reconstruction methods falls off rapidly as burst length increases. Interleaving, feature estimation and decoder based strategies are combined in [36] in order to improve the recognition performance in the presence of packet loss in DSR.

In this section, the 3-state model proposed in [33] is used to simulate packet loss and loss bursts. To compensate for missing packets, two error-concealment methods are examined, namely nearest neighbour repetition and interpolation. Error mitigation using interleaving is also considered.

4.2. Packet Loss Framework

4.2.1. Packet Loss Model

The packet loss model used in this work is the 3-state Markov chain proposed by [33]. This 3-state model was found to be more effective at simulating different packet loss conditions in comparison with a 2-state Markov chain and the Gilbert-Elliot model. The model is detailed in Figure 3, showing the three states and the transition probabilities. Occupancy of states 1 and 3 indicate no packet loss while occupancy of state 2 indicates packet loss. In Figure 3, Q, q and Q' are the self-loop probabilities of states 1, 2 and 3 respectively. The model parameters are designed so that state 1 models long duration periods of no loss and state 3 models short periods of no loss, which occur in between packet loss in burst-like conditions. The following four parameters define the model, and from these parameters the transition probabilities of the 3-state model can be determined:

α = overall probability of a packet being lost

β = average packet loss burst length

N_1 = average length, in packets, of loss-free periods

N_3 = average length, in packets, of no-loss periods inside loss periods

The transition probabilities are calculated from the following equations:

$$q = 1 - \frac{1}{\beta} \tag{2}$$

$$Q = 1 - \frac{1}{N_1} \tag{3}$$

$$Q' = 1 - \frac{1}{N_3} \tag{4}$$

$$p = \frac{1-Q}{Q-Q'}\left[\frac{1-Q'}{\alpha} + q + Q' - 2\right] \tag{5}$$

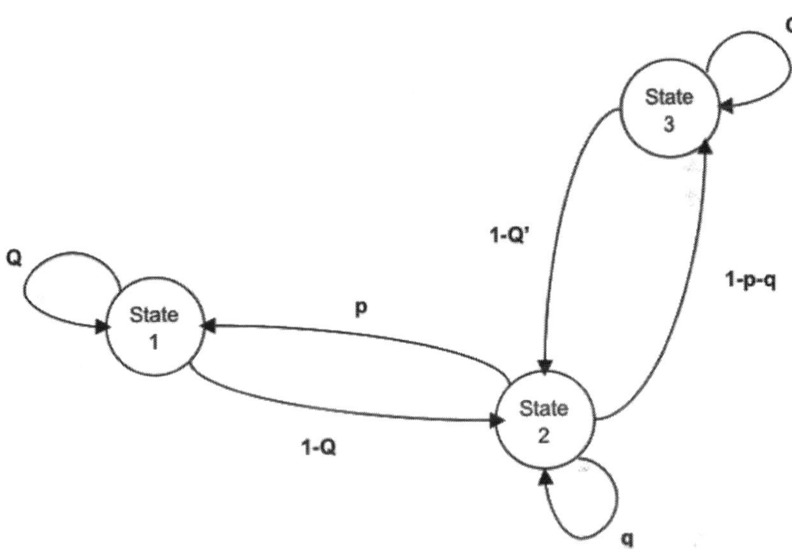

Figure 3. Packet Loss Model [**33**].

The authors in [33] suggest that an alternative to performing speech recognition tests using simulated channels is to define a set of packet loss characteristics, thus enabling recognition performance to be analysed across a range of different packet loss conditions. References [34-37] define four channels with different characteristics in order to simulate packet loss. These same four channels are used here to determine the effect of packet loss on speech recognition performance. The parameter values for the four channels are detailed in Table 6. These parameters result from work in [33] on IP and wireless networks. These are network environments over which a DSR system would typically operate. In an IP network, packet loss arises primarily due to congestion at the routers within the network, due to high levels of IP traffic. The nature of IP traffic is that it can be described as being 'bursty' in nature with the result that packet loss occurs in bursts. Signal fading, where the signal strength at a receiving device is attenuated significantly, is also a contributing factor to packet loss in a wireless network. Long periods of fading in a wireless network can result in bursts of packet loss. The authors in [33] measured the characteristics of an IP network and a WLAN, and the results showed the packet loss rate (α) and the burst length (β) to be highly variable. At one point or another, most channel conditions occurred, although not necessarily for long. Based on the experimental measurements, a set of packet loss characteristics was defined in [33] and these are used to analyse recognition performance for different network conditions. The parameters in Table 6 are taken from this defined set of packet loss characteristics.

Table 6. Packet loss parameters.

	α	β	N_1	N_3
Channel A	10%	4	37	1
Channel B	10%	20	181	1
Channel C	50%	4	5	1
Channel D	50%	20	21	1

4.2.2. Packet Loss Mitigation

Two error concealment methods are examined, namely nearest neighbour repetition and interpolation. These methods attempt to reconstruct the feature vector stream when packet loss is detected. Missing feature vectors are estimated solely from correctly received feature vectors. In a DSR system, nearest neighbour repetition and interpolation would both be implemented on the server side. Additionally, interleaving, a technique used to reduce feature vector loss burst lengths (with a penalty of additional delay), is also briefly discussed. Interleaving is carried out on the client side of a DSR system, with de-interleaving on the server side.

4.2.2.1. Nearest neighbour repetition

The ETSI advanced front-end [7] specifies that where missing feature vectors occur due to transmission errors, they should be substituted with the nearest correctly received feature vector in the receiver. If there are 2*B* consecutive missing feature vectors, the first *B* speech vectors are substituted by a copy of the last good speech vector before the error, and the last *B* speech vectors are substituted by a copy of the first good speech vector received after the error. The speech vector includes the 12 static cepstral coefficients C_1-C_{12}, the zeroth cepstral coefficient C_0 and the log energy term, and all are replaced together. A disadvantage of this method is that if *B* is large then long stationary periods can arise.

4.2.2.2. Interpolation

The disadvantage of stationary periods that arise with nearest neighbour repetition can be alleviated somewhat by polynomial interpolation between the correctly received feature vectors either side of a loss burst. Reference [34] found that non-linear interpolation using cubic Hermite polynomials gives the best estimates for missing feature vectors. Equation (6) is used to calculate the nth missing feature vector in a loss burst of length *β* packets, which is equivalent to a loss burst length of 2*β* feature vectors if each packet contains two feature vectors as defined by the ETSI advanced front-end [7]. The parameter *n* in (6) is the missing feature vector index.

$$\hat{x}_{b+n} = a_0 + a_1\left(\frac{n}{\beta+1}\right) + a_2\left(\frac{n}{\beta+1}\right)^2 + a_3\left(\frac{n}{\beta+1}\right)^3 \quad 1 \le n \le 2\beta \tag{6}$$

The coefficients 0a , 1a , 2 a and 3 a in (6) are determined from the two correctly received feature vectors either side of the loss burst, b x and b n 1 x + + , and their first derivatives, b x′ and b n 1 x + + ′ . Equation (6) can be rewritten as

$$\hat{x}_{b+n} = x_b\left(1-3t^2+2t^3\right) + x_{b+\beta+1}\left(3t^2-2t^3\right) + x'_b\left(t-2t^2+t^3\right) + x'_{b+\beta+1}\left(t^3-t^2\right) \quad 1 \le n \le 2\beta \tag{7}$$

where $t = n/(\beta+1)$.. It was found in [34] that performance was better when the derivative components in (7) are set to zero. These components are also set to zero for the work presented in this chapter.

4.2.2.3. Interleaving

Research has shown that by minimising the average burst length of lost vectors the estimation of lost feature vectors is more effective [34]. The aim of interleaving is to break a long loss burst into smaller loss bursts by distributing them over time and so making it appear that the errors are more randomly distributed.

In a DSR system, the interleaver on the client side takes a feature vector sequence X_i, where i is the order index, and changes the order in which the vectors are transmitted over the channel. The result is to generate a new vector sequence Y_i that is related to X_i by

$$Y_i = X_{\pi(i)} \tag{8}$$

where $\pi(i)$ is the permutation function. On the server side, the operation is reversed by de-interleaving the received vector sequence as follows:

$$X_i = Y_{\pi^{-1}(i)} \tag{9}$$

where $\pi(\pi^{-1}(i))=i$.

In order for the interleaver to carry out the reordering of the feature vectors, it is necessary to buffer the vectors, which introduces a delay. On the server side, in order to carry out the de-interleaving, buffering of the incoming feature vectors takes place and a second delay is introduced. The sum of these two delays is known as the latency of the interleaving/de-interleaving process.

The spread S of an interleaver is a metric that indicates how good an interleaver is at breaking up error bursts. A burst of length L vectors will be broken into bursts of length 1 if $S \geq L$. For $S < L$ the full distribution of the burst cannot be guaranteed and some sets of consecutive feature vectors may be lost.

For the work in this chapter, block interleaving is implemented. A block interleaver of degree *d*changes the order of transmission of a *dxd* block of input vectors. An example of a block interleaver of degree $d = 4$ and spread $S = 4$ is given in Figure 4.

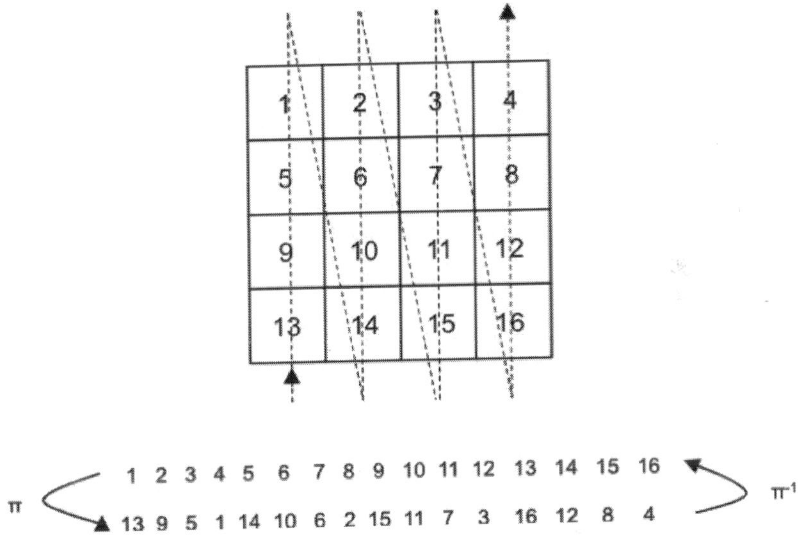

Figure 4. dxd block interleaver where d = 4, with permutation function.

4.3. Tests and Results

The primary purpose of this section is to investigate the performance of the auditory model proposed by Li *et al.* [14] in combination with speech enhancement in the presence of noise and packet loss. As a baseline for comparison, results are also presented for the ETSI advanced front-end [7]. In all cases, training was carried out using clean data. The speech enhancement algorithm of Agarwal and Cheng [25] was used on both the (clean) training speech as well as the (noisy) test speech. The ETSI advanced front-end includes a SNR-dependent waveform processing block that is applied after noise reduction and before feature extraction. The waveform processing block in the ETSI advanced front-end is also implemented in the front-end of Li *et al.* in order to ensure a closer match between the two front-ends. Feature vectors are extracted from the output of this waveform processing block. A detailed description of the waveform processing block can be found in [26]. The ETSI advanced front-end carries out post-processing in the cepstral domain in the form of blind equalization as described by [13]. The feature vectors produced by Li *et al.* undergo post-processing in the cepstral domain by means of cepstral mean subtraction (CMS). As defined by the ETSI advanced front-end [7], each packet transmitted over the communication channel carries two feature vectors.

In [7] a distributed speech recognition front-end feature vector compression algorithm is defined. The algorithm makes use of the parameters from the front-end feature extraction algorithm of the ETSI advanced front-end. The purpose of the algorithm is to reduce the number of bits needed to represent each front-end feature vector and so reduce the bit rate required over the communications channel. The feature vector is directly quantized with a split vector quantiser. The 14 coefficients (C_1 to C_{12}, C_0 & lnE) are grouped into pairs, and each pair is quantized using its own vector quantisation (VQ) codebook. The resulting set of index values is then used to represent the feature vector. The results documented in this paper are based on feature vectors that have undergone split vector quantisation.

The baseline recognition results for the two front-ends, without vector quantisation and with no packet loss but with noise, are detailed in Table 7. The word accuracies in the following tables are calculated as described in Section 2.4.

Table 7. Baseline recognition results.

	Absolute word accuracy %			
Front-end	Set A	Set B	Set C	Overall
ETSI AFE	87.74	87.09	85.44	87.02
Li *et al.*	88.62	88.09	86.89	88.06

In order to implement split vector quantization it is necessary to design VQ codebooks for each of the seven coefficient pairs. ETSI has made available script files for the ETSI advanced front-end and included with these are the VQ codebooks for the coefficient pairs. The recognition results for the ETSI advanced front-end with feature vector quantization using the ETSI supplied VQ codebooks are given in Table 8.

To allow for close comparison between the ETSI advanced front end and the front-end proposed by Li*et al.*, the VQ codebooks for Li *et al.* should be determined in the same manner as the VQ codebooks for the ETSI advanced front-end. However, this was not possible as the detail of how the ETSI advanced front-end VQ codebooks were calculated is not publicly available at this time. Therefore, an implementation of the Generalized Lloyd Algorithm (GLA), described by [39], was used to design the VQ codebooks for both the ETSI advanced front-end and the front-end of Li *et al.* The recognition results for the two front-ends using the VQ codebooks generated by the GLA implementation are detailed in Table 9. The overall word accuracies in Table 9, with vector quantization, compare well with the baseline accuracies, without vector quantization, in Table 7. There is also close correlation between the recognition results in Table 8 and Table 9 for the ETSI advanced front-end, indicating that the VQ codebooks generated by the GLA implementation used for this work are a good substitute for the VQ codebooks provided by ETSI with the advanced front-end.

Table 8. Recognition results using ETSI VQ codebooks.

	Absolute word accuracy %			
Front-end	Set A	Set B	Set C	Overall
ETSI AFE	87.81	87.11	85.74	87.12

Table 9. Recognition results with VQ codebooks designed using implementation of the GLA.

	Absolute word accuracy %			
Front-end	Set A	Set B	Set C	Overall
ETSI AFE	87.73	86.92	85.41	86.94
Li *et al.*	88.22	87.59	86.55	87.63

Packet loss (where each packet contains two feature vectors) is introduced on the communication channel by using the packet loss model described in Section 4.2.1. The four different packet loss channels investigated are defined in Table 6. Recognition tests, in the presence of packet loss, were carried out for each of the following conditions:

- no speech enhancement, no loss mitigation (Table 10);
- speech enhancement, no loss mitigation (Table 11);
- speech enhancement, nearest neighbour repetition (Table 12);
- speech enhancement, interpolation (Table 13);
- speech enhancement, interleaving, interpolation (Table 14).

Tests were first carried out for packet loss with no steps taken to recover the missing features or to minimise the loss burst length. The test results for both front-ends when no speech enhancement is used are given in Table 10, while recognition results with speech enhancement are given in Table 11. A comparison of Table 10 with Table 11 illustrates the benefit of using speech enhancement in improving recognition performance. Comparing Table 11 with Table 9 (no packet loss) it is seen that packet loss has a significant impact on the recognition results, in particular for channels C and D where the probability of packet loss is 50%.

Table 10. No speech enhancement, no error mitigation.

	ETSI AFE absolute word accuracy %			
Loss parameters	Set A	Set B	Set C	Overall
Channel A	60.56	60.96	64.36	61.48
Channel B	59.31	59.49	63.67	60.25
Channel C	36.45	37.62	37.66	37.16
Channel D	35.38	35.89	37.68	36.04
	Li *et al.* absolute word accuracy %			
Loss parameters	Set A	Set B	Set C	Overall
Channel A	66.66	68.39	63.85	66.79
Channel B	65.59	67.35	62.93	65.76
Channel C	38.27	39.60	36.29	38.41
Channel D	37.78	39.25	36.07	38.03

Table 11. Speech enhancement, no error mitigation.

	ETSI AFE absolute word accuracy %			
Loss parameters	Set A	Set B	Set C	Overall
Channel A	80.17	80.09	77.97	79.70
Channel B	79.86	79.28	77.61	79.18
Channel C	42.50	42.87	40.50	42.25
Channel D	44.07	44.25	42.21	43.77
	Li *et al.* absolute word accuracy %			
Loss parameters	Set A	Set B	Set C	Overall
Channel A	81.08	80.64	79.37	80.56
Channel B	80.33	79.85	78.85	79.84
Channel C	43.35	43.53	41.91	43.13
Channel D	44.46	44.30	43.44	44.19

Two methods, nearest neighbour repetition and Hermite interpolation, are used to reconstruct the feature vector stream as a result of missing features due to packet loss. Table 12 details the recognition results obtained when using nearest neighbour repetition while Table 13 details the results obtained when Hermite interpolation is implemented (speech enhancement is used in both cases). Both reconstruction methods show improvements in recognition testing over no error mitigation for all four channels. In particular, with feature reconstruction channel C shows improvements in recognition accuracy greater than 55% for both front-ends. Channel D also shows good improvement. Nearest neighbour repetition gives a slightly higher performance compared to Hermite interpolation.

When interleaving is introduced, the receive side perceives that the average loss burst length is reduced [37]. Table 14 shows the recognition results obtained when interleaving, with an interleaving depth of 4, is used in conjunction with Hermite interpolation. Comparing the results in Table 14 with the results in Table 13 it is seen that feature reconstruction is improved when interleaving is employed.

Table 12. Speech enhancement, nearest neighbour repetition.

	ETSI AFE absolute word accuracy %			
Loss parameters	Set A	Set B	Set C	Overall
Channel A	84.04	83.34	81.69	83.29
Channel B	80.85	80.37	78.89	80.26
Channel C	68.90	68.25	66.83	68.23
Channel D	50.95	50.82	50.67	50.84
	Li *et al.* absolute word accuracy %			
Loss parameters	Set A	Set B	Set C	Overall
Channel A	84.57	84.00	82.67	83.96
Channel B	81.10	80.89	79.78	80.75
Channel C	68.81	68.60	66.87	68.34
Channel D	50.33	51.10	49.86	50.54

Table 13. Speech enhancement, Hermite interpolation.

	ETSI AFE absolute word accuracy %			
Loss parameters	Set A	Set B	Set C	Overall
Channel A	83.87	83.28	82.01	83.26
Channel B	80.81	80.21	78.78	80.17
Channel C	67.62	68.03	66.28	67.52
Channel D	50.33	50.16	48.91	49.98
	Li *et al.* absolute word accuracy %			
Loss parameters	Set A	Set B	Set C	Overall
Channel A	84.23	83.69	82.56	83.68
Channel B	80.87	80.60	79.64	80.51
Channel C	67.05	67.03	65.32	66.70
Channel D	50.33	50.16	48.91	49.98

Table 14. Speech enhancement, Hermite interpolation with interleaving (*d* = 4).

	ETSI AFE absolute word accuracy %			
Loss parameters	Set A	Set B	Set C	Overall
Channel A	86.59	85.66	84.25	85.75
Channel B	82.65	81.89	80.51	81.92
Channel C	78.35	77.92	76.67	77.84
Channel D	58.52	58.77	59.88	58.89
	Li *et al.* absolute word accuracy %			
Loss parameters	Set A	Set B	Set C	Overall
Channel A	86.91	86.29	85.20	86.32
Channel B	82.84	82.15	81.50	82.30
Channel C	78.14	77.92	76.73	77.77
Channel D	57.95	58.13	57.05	57.84

4.4. Discussion

The results in Table 7 show that the front-end proposed by Li et al. [14], when combined with the speech enhancement algorithm proposed by [25], reduces the overall word error rate of the ETSI advanced front-end [7] by 8%. Looking at Table 9, the vector quantisation has a lesser impact on the overall recognition performance of the ETSI advanced front-end compared with the impact of vector quantisation on the Li et al. front end. The Li et al. front-end, combined with speech enhancement, still outperforms the ETSI advanced front-end in the presence of vector quantisation although the improvement in overall word error rate is reduced from 8% (without vector quantisation) to 5.3%. In the presence of packet loss, with no speech enhancement and with no packet loss compensation, a comparison of Table 10 shows that the front-end of Li et al. gives better overall recognition results than the ETSI advanced front-end. The benefit of speech enhancement in the presence of packet loss, without any missing feature reconstruction, can be seen by comparing Table 10 and Table 11. With speech enhancement and no packet loss compensation, Table 11 shows that Li et al. outperforms the ESTI advanced front-end for all four channels, and for all three test sets. Comparing Table 9 withTable 11, a significant reduction in recognition performance is observed in the presence of packet loss, in particular for channels C and D where the probability of packet loss is 50%. When nearest neighbour repetition is used to reconstruct missing features, Table 12 shows that there is a significant increase in recognition performance across all channels when compared to the results presented inTable 11. Looking at Table 12, the recognition results for the two front-ends under evaluation are similar across all channels and test sets. The front-end of Li et al. performs marginally better overall than the ETSI advanced front-end for channels A, B and C; however, for channel D, the overall recognition performance of the ETSI advanced front-end is better than that of Li et al. A comparison of Table 13 with Table 12 shows a slight decrease in recognition performance when Hermite interpolation is used to reconstruct the feature vector stream instead of nearest neighbour repetition. With Hermite interpolation, Table 13 shows that the front-end of Li et al. outperforms the ETSI advanced front-end for the packet loss conditions of channels A and B, however, for channel C the reverse is the case. The overall performance of both front-ends is the same for channel D. Interleaving the feature vectors prior to transmission on the channel gives the perception on the receive side that the loss bursts are shorter than they actually are. The advantage of interleaving can be seen by a comparison of Table 14 with Table 13, where overall recognition results are improved for both front-ends when interleaving is introduced. Looking at Table 14 it is seen that Li et al. gives the better overall recognition performance for channels A and B while the ETSI advanced front-

5. CONCLUSIONS

This chapter has examined the speech recognition performance of both a speech enhancement algorithm combined with the auditory model front-end proposed by Li et al. [14], and the ETSI advanced front-end [7], in the presence of both environmental noise and packet loss. A number of speech enhancement techniques were first examined, including well-established techniques such as Ephraim and Malah [20] and more recently-proposed techniques such as Rangachari and Loizou [21]. Experiments using the Aurora connected-digit recognition framework [16] found that the best performance was obtained using the method of Agarwal and Chang [25]. The test results also suggest that the choice of speech enhancement algorithm for best speech recognition performance is independent of the choice of front-end.

Packet loss modelling using statistical modelling was also examined, and packet loss mitigation was discussed. Following initial testing with no packet loss compensation, a number of existing packet loss mitigation techniques were investigated, namely nearest neighbour repetition and interpolation. Results show that the best recognition performance was obtained using nearest neighbour repetition to reconstruct missing features. The advantage of interleaving at the sender's side to minimise the average burst length of lost vectors was also demonstrated.

In summary, the experiments and results outlined in this chapter show the benefit of combining speech enhancement and packet loss mitigation to combat both noise and packet loss. Furthermore, the performance of the auditory model of Li et al. was generally shown to be superior to that of the standard ETSI advanced front-end.

REFERENCES

1. V. Digalakis, L. Neumeyer and M. Perakakis, "Quantization of cepstral parameters for speech communication over the world wide web", IEEE Journal on Selected Areas in Communications, vol. 17, pp. 82-90, Jan. 1999.

2. D. Pearce, "Enabling new speech driven services for mobile devices: an overview of the ETSI standards activities for distributed speech recognition front-ends", in Proc. AVIOS 2000: The Speech Applications Conference, San Jose, CA, USA, May 2000.

3. G. Gallardo-Antolín, C. Peláez-Moreno and F. Díaz-de-María, "Recognizing GSM digital speech", IEEE Trans. on Speech and Audio Processing, vol. 13, pp. 1186-1205, Nov. 2005.

4. Z.H. Tan, P. Dalsgaard and B. Lindberg, "Automatic speech recognition over error-prone wireless networks", Speech Communication, vol. 47, pp. 220-242, 2005.

5. Y. Ephraim and I. Cohen, "Recent advancements in speech enhancement", in The Electrical Engineering Handbook, 3rd ed., R.C. Dorf, Ed., Boca Raton, FL: CRC Press, 2006, pp. 15-12 to 15-26.

6. "Speech Processing, Transmission and Quality Aspects (STQ); Distributed speech recognition; Front-end feature extraction algorithm; Compression algorithms", in ETSI ES 201 108, Ver. 1.1.3, Sept. 2003.

7. “Speech Processing, Transmission and Quality Aspects (STQ); Distributed speech recognition; Advanced front-end feature extraction algorithm; Compression algorithms”, in ETSI ES 202 050, Ver. 1.1.5, Jan. 2007.

8. “Speech Processing, Transmission and Quality Aspects (STQ); Distributed speech recognition; Extended front-end feature extraction algorithm; Compression algorithms; Back-end speech reconstruction algorithm”, in ETSI ES 202 211, Ver. 1.1.1, Nov. 2003.

9. “Speech Processing, Transmission and Quality Aspects (STQ); Distributed speech recognition; Extended advanced front-end feature extraction algorithm; Compression algorithms; Back-end speech reconstruction algorithm”, in ETSI ES 202 212, Ver. 1.1.2, Nov. 2005.

10. “RTP Payload Format for European Telecommunications Standards Institute (ETSI) European Standard ES 210 108 Distributed Speech Recognition Encoding”, in RFC 3557, July 2003.

11. “RTP Payload Formats for European Telecommunications Standards Institute (ETSI) European Standard ES 202 050, ES 202 211, and 202 212 Distributed Speech Recognition Encoding”, in RFC 4060, May 2005.

12. D. Macho, L. Mauuary, B. Noe, Y. M. Cheng, D. Ealey, D. Jouvet, H. Kelleher, D. Perace, and F. Saadoun, “Evaluation of a noise-robust DSR front-end on Aurora databases”, in Proceedings of International Conference on Speech and Language Processing, Denver, Colorado, USA, Sept. 2002, pp. 17-20.

13. L. Mauuary, “Blind equalization in the cepstral domain for robust telephone based speech recognition”, in Proc. EUSIPCO 98, Sept. 1998, pp. 359-363.

14. Q. Li, F. K. Soong and O. Siohan, “A high-performance auditory feature for robust speech recognition”, in Proc. of 6th International Conference on Spoken Language Processing (ICSLP), Beijing, China, Oct. 2000, pp. 51-54.

15. R. Flynn and E. Jones, “A comparative study of auditory-based front-ends for robust speech recognition using the Aurora 2 database”, in Proc. IET Irish Signals and Systems Conference, Dublin, Ireland, 28-30 June 2006, pp. 111-116.

16. H. G. Hirsch and D. Pearce, "The Aurora experimental framework for the performance evaluation of speech recognition systems under noisy conditions", in Proc. ISCA ITRW ASR-2000, Paris, France, Sept. 2000, pp. 181-188.

17. H. Hermansky, "Perceptual linear prediction (PLP) analysis of speech", J. Acoust. Soc. Amer., vol. 87, pp. 1738-1752, 1990.

18. J. Tchorz and B. Kollmeier, "A model of auditory perception as front end for automatic speech recognition", J. Acoust. Soc. Amer., vol. 106, pp. 2040-2050, 1999.

19. HTK speech recognition toolkit. Available: http://htk.eng.cam.ac.uk/. Accessed March 2011.

20. Y. Ephraim and D. Malah, "Speech enhancement using a minimum mean-square error short-time spectral amplitude estimator", IEEE Trans. on Acoustics, Speech and Signal Processing, vol. 32, pp. 1109-1121, Dec. 1984.

21. S. Rangachari and P. Loizou, "A noise-estimation algorithm for highly non-stationary environments", Speech Communication, vol. 48, pp. 220-231, 2006.

22. N. Westerlund, M. Dahl and I. Claesson, "Speech enhancement for personal communication using an adaptive gain equalizer", Speech Communication, vol. 85, pp. 1089-1101, 2005.

23. R. Martin, "Spectral subtraction based on minimum statistics", in Proc. Eur. Signal Processing Conference, 1994, pp. 1182-1185.

24. R. Martin, "Noise power spectral density estimation based on optimal smoothing and minimum statistics", IEEE Trans. on Speech and Audio Processing, vol. 9, pp. 504-512, July 2001.

25. A. Agarwal and Y.M. Cheng, "Two-stage mel-warped wiener filter for robust speech recognition", in Proceedings of Automatic Speech Recognition and Understanding Workshop, Keystone, Colorado, USA, 1999, pp. 67-70.

26. D. Macho and Y. M. Cheng, "SNR-dependent waveform processing for improving the robustness of ASR front-end", in Proc. of IEEE

International Conference on Acoustics, Speech and Signal Processing (ICASSP), May 2001, pp. 305-308.

27. C. Peláez-Moreno, G. Gallardo-Antolín and F. Díaz-de-María, "Recognizing voice over IP: A robust front-end for speech recognition on the world wide web", IEEE Trans. on Multimedia, vol. 3, pp. 209-218, June 2001.

28. C. Peláez-Moreno, G. Gallardo-Antolín, D.F. Gómez-Cajas and F. Díaz-de-María, "A comparison of front-ends for bitstream-based ASR over IP", Signal Processing, vol. 86, pp. 1502-1508, July 2006.

29. J. Van Sciver, J. Z. Ma, F. Vanpoucke and H. Van Hamme, "Investigation of speech recognition over IP channels", in Proc. of IEEE International Conference on Acoustics, Speech and Signal Processing (ICASSP), May 2002, pp. 3812-3815.

30. D. Quercia, L. Docio-Fernandez, C. Garcia-Mateo, L. Farinetti and J. C. DeMartin, "Performance analysis of distributed speech recognition over IP networks on the AURORA database", in Proc. of IEEE International Conference on Acoustics, Speech and Signal Processing (ICASSP), May 2002, pp. 3820-3823.

31. P. Mayorga, L. Besacier, R. Lamy and J. F. Serignat, "Audio packet loss over IP and speech recognition", in Proc. of IEEE Workshop on Automatic Speech Recognition and Understanding (ASRU), Nov. 2003, pp. 607-612.

32. J. Vicente-Peña, G. Gallardo-Antolín, C. Peláez-Moreno and F. Díaz-de-María, "Band-pass filtering of the time sequences of spectral parameters for robust wireless speech recognition", Speech Communication, vol. 48, pp. 1379-1398, Oct. 2006.

33. B.P. Milner and A.B. James, "An analysis of packet loss models for distributed speech recognition", in Proc. of 8th International Conference on Spoken Language Processing (ICSLP), Jeju Island, Korea, Oct. 2004, pp. 1549-1552.

34. A. B. James and B. P. Milner, "An analysis of interleavers for robust speech recognition in burst-like packet loss", in Proc. of IEEE International Conference on Acoustics, Speech and Signal Processing (ICASSP), May 2004, pp. 853-856.

35. A. B. James and B. P. Milner, "Towards improving the robustness of distributed speech recognition in packet loss", in Proc. Second COST278 and ISCA Tutorial and Research Workshop (ITRW) on Robustness Issues in Conversational Interaction, University of East Anglia, U.K., Aug. 2004, p. paper 42.

36. A. B. James and B. P. Milner, "Combining packet loss compensation methods for robust distributed speech recognition", in Proc. of Interspeech-2005, Lisbon, Portugal, Sept. 2005, pp. 2857-2860.

37. A. B. James and B. P. Milner, "Towards improving the robustness of distributed speech recognition in packet loss", Speech Communication, vol. 48, pp. 1402-1421, Nov. 2006.

38. B. P. Milner and A. B. James, "Robust speech recognition over mobile and IP networks in burst-like packet loss", in IEEE Trans. on Audio, Speech and Language Processing vol. 14, ed, Jan. 2006, pp. 223-231.

39. Y. Linde, A. Buzo and R. Gray, "An algorithm for vector quantizer design", IEEE Trans. on Communications, vol. COM-28, pp. 84-95, Jan. 1980.

Chapter 3

Advanced acoustic modelling techniques in MP3 speech recognition

Michal Borsky*, Petr Pollak and Petr Mizera

Department of Circuit Theory, Czech Technical University, Technická 2, 166 27, Prague 6, Czech Republic

ABSTRACT

The automatic recognition of MP3 compressed speech presents a challenge to the current systems due to the lossy nature of compression which causes irreversible degradation of the speech wave. This article evaluates the performance of a recognition system optimized for MP3 compressed speech with current state-of-the-art acoustic modelling techniques and one specific front-end compensation method. The article concentrates on acoustic model adaptation, discriminative training, and additional dithering as prominent means of compensating for the described distortion in the task of phoneme and large vocabulary continuous speech recognition (LVCSR). The experiments presented on the phoneme task show a dramatic increase of the recognition error for unvoiced speech units as a direct result of compression. The application of acoustic model adaptation has proved to yield the highest relative contribution while the gain of discriminative training diminished with decreasing bit-rate. The application of additional dithering yielded a consistent improvement only for the MFCC features, but the overall results were still worse than those for the PLP features.

KEYWORDS

MP3 compression, Phoneme recognition, LVCSR, GMM-HMM, Acoustic modelling, Additional dithering, AM adaptation, Discriminative training

1. INTRODUCTION

The aim of automatic speech recognition (ASR) research is to develop an intermediary system for the purpose of human speech transcription where the construction and block architecture is often customized. The transcription of digitally stored data represents an example where the ASR systems has to be specifically tailored to perform optimally. Ordinary people, call centers, and media companies have large amounts of data stored for the purpose of further processing or simply for later accessibility. This data can consume large amounts of storage capacity. The obvious solution to this problem has been to compress these recordings using an audio coder with high compression rate. Although it may initially have been assumed that these compressed recordings would be accessed by a human listener, the introduction of automatic speech processing systems has changed this paradigm.

MPEG-2 AudioLayer III, also known as MP3, belongs to the group of perceptual audio codecs, which are based on the physiology of human hearing. Its main advantage is a relatively high compression rate while at the same time retaining good intelligibility for human listeners. This characteristic is the reason for its wide-spread use in the personal, commercial, and public sphere. On the other hand, the compression introduces severe distortions which limit its use for audio professionals or for automatic speech recognition. The application of auditory masking functions and the quantization in the compression scheme results in speech distortion. The exact nature of the distortion has been studied in [1] and [2]. The two main problems identified by the authors were the bandwidth limitation and spectral valleys.

This article investigates the performance of current state-of-the-art acoustic modelling (AM) and feature extraction techniques in the task of phoneme and large vocabulary continuous speech recognition of MP3 compressed speech. It is organized as follows: the next section gives a short overview of related works on this topic, followed by a theoretical analysis of distortion for spectral-based speech features, a description of used techniques, and a section detailing the experimental setup and achieved results. The article concludes with a discussion.

2. MP3 SPEECH RECOGNITION

Studies on practical usability of MP3 recordings in automatic speech recognition concluded that the system can perform without degradation for sufficiently high bit-rates. The authors consistently reported a significant drop in accuracy for bit-rates lower than 24 kbps, i.e., [3, 4, 5, 6, 7]. Several solutions have been proposed to improve the recognition for lower bit-rates, starting with limiting the training signal bandwidth, using perceptual linear prediction (PLP) features or adding a controlled amount of noise.

2.1 Related works

The MP3 format actively limits the spectral bandwidth of the compressed data, which can create the problem of training and testing data mismatch. To solve this problem, and to avoid compressing and decompressing the whole training subset, the authors in [3] proposed a parameterization scheme with a low-pass filter in the block of the front-end processing. A specific cutoff frequency was assigned to each bit-rate, and the AMs were trained on the filtered speech and then tested on the compressed speech. The method yielded only a marginal decrease in word error rate (*WER*) by 1–2 % in a simple digit recognition task, depending on the bit-rate. The second problem of bandwidth limitation is the loss of information carried by higher frequencies. This is expected to mainly affect the speech units without strong low-frequency harmonic structure, such as unvoiced consonants, and subsequently result in increased likelihood of false recognition of these units.

In [7], the authors studied the effect of spectral valleys on speech features and concluded that the main portion of degradation could be attributed to the spectral valleys. The valleys act as a step energy change between neighboring frames, which increases the values of Δ and Δ^2 parameters and randomly displaces their position in the feature space. The proposed solution was based on adding a controlled amount of noise to "fill in"? the valleys and to reduce the features' variance. The report demonstrated that the application of this technique could bring significant WER reduction by up to 45 % for low bit-rates and Mel-frequency cepstral coefficient (MFCC) features. In general, better results were obtained for lower bit-rates while the results for higher bit-rates were only slightly degraded due to the introduction of additional noise.

A comparative study of spectral-based features for MP3 speech recognition [6] demonstrated the advantage of using PLPs over MFCCs where the PLP features outperformed the MFCCs by 11 % absolute at a bit-rate of 24 kbps. The reported WER differences for higher bit-rates were much lower, at 4–6 %. The authors concluded that this behavior could be attributed to the application of the equal loudness curve and the psychoacoustic scaling of the analysis filter bank.

2.2 Robust front-end processing

Cepstral mean normalization (CMN) is a well-established technique for robust speech recognition. The principle is based on the assumption that if the convolutional noise in a short time-frame is stationary, then it can be subtracted from the extracted features in the logarithmic spectral domain. Although it is a fairly simple method, it has been proven to provide robustness against environmental and channel distortions and speaker variability.

Linear discriminant analysis (LDA) aims to improve separability among classes by transforming the feature vector of dimension *n* to the vector of dimension *m*. It is typically used in ASR systems to reduce the dimensionality of spliced feature vectors and to decorrelate the features. However, it has also been shown to improve the performance of standard features in the presence of noise [8].

Modifications of the signal can occasionally result in a sequence of zeros in the time domain which, if not treated properly, can cause the extraction algorithm to fail. If the standard procedure to avoid the *Inf.* values in logarithmic spectra is to add small amounts of uniformly distributed noise, then the addition of relatively strong noise has been shown to improve the recognition of spectrally distorted speech [7]. This technique is referenced as additional dithering later in the text.

2.3 MP3 acoustic modelling

Since MP3 recognition failure is primarily influenced by the changes at the feature extraction and acoustic modelling levels, additional refinements are required to compensate for the decline. This section provides an overview of methods used in a situation with adverse conditions or, in general, in order to increase the quality of AM.

The AM adaptation has been documented to perform well in situations with a training and testing data mismatch [9,10] when the commonly used approach is to adapt the existing AM to the new conditions. The technique employed in this article was the feature maximum likelihood linear regression (fMLLR). Its main advantage is its robustness against the lack of adaptation data and the ability to estimate new model parameters even from the erroneous transcription.

The conventional system based on Gaussian mixture models can contain several hundred thousand of mixtures. A subspace Gaussian mixture model (SGMM) has been proposed as an alternative approach in which the model parameters are typically initialized from a clustered model, i.e., universal background model, and then retrained and shared among multiple models. The result is a reduction in model parameters, which allows estimation of SGMM parameters using a smaller amount of data and is expected to better model the acoustic variabilities.

Discriminative training has become a common modelling method when the main principle is based on formulating an objective function tied to the classification and minimizing the recognition error directly instead of maximizing the observation likelihood. Some published works have reported on its usage for noisy speech recognition, e.g., [11] or [12], and concluded that its usage can increase the robustness of the system. The main drawback of discriminative training may be the lack of generalization to the test set, which is likely to occur if the uncompressed, discriminatively trained AM is deployed to recognize the compressed speech. Despite this obvious disadvantage, it was expected that the modelling improvement would outweigh the generalization problem for our task.

3. EXPERIMENTAL EVALUATION

This section describes a series of experiments investigating the influence of various acoustic modelling and modified feature extraction techniques in the task of MP3 recognition. The experiments were performed using the Kaldi [13] toolkit, and the described recognition system was based on the Gaussian mixture model-hidden Markov model (GMM-HMM) approach.

3.1 Common experimental setup

The signals for experiments came from the Czech SPEECON and TEMIC database, were recorded in 16-bit precision and 16-kHz sampling frequency by a headset microphone, and were manually transcribed. The data were split randomly into train and test subsets. The MP3 compression for the test subset was simulated by Lame [14] software, and SOX was used for the decompression. The compression rates were selected with the intention of evaluating the performance of the system for bit-rates of 128, 32, 28, 24, 20, 16, and 12 kbps.

The 39-dimensional PLP and MFCC features were computed using the CtuCopy extraction tool [15] with 32 ms window and 16 ms shift. The CMN technique was applied in a speaker-specific fashion and on static features only. In the first stage of the experiments, the signals were dithered with uniformly distributed random values from the <−1,1> range. The effect of additional dithering for the test subset was studied in the later stages, when the dithering value $<-R,R>$ was gradually increased.

The AMs were trained on uncompressed speech using the Viterbi training algorithm from 72 h of speech and 555 speakers. The baseline uncompressed system consisted of continuous density hidden Markov models for context-dependent crossword triphones. The basic phonetic alphabet contained 44 Czech monophones and silence. The quality of the baseline AM was later improved by LDA, speaker adaptive training (SAT), SGMM framework, and discriminative training (DT). The initial feature vector for LDA was extended by three neighboring vectors, and its dimensionality was then reduced to 40. The fMLLR adaptation was used in the SAT scheme to produce the speaker-independent AMs. The final AMs were adapted in an unsupervised, speaker-specific fashion, and the maximum mutual information (MMI) criterion was used for DT. The weighted finite state machine decoder was used for recognition.

3.2 Results of optimized acoustic modelling

The preparation of the test subset involved compression and decompression from which the features were extracted afterwards. The phoneme test subset contained 45 speakers and 8.5 h of speech of varying content, and the recognition was performed using a bigram phoneme model. The results were evaluated by phone error rate (*PER*) and the phone error rate reduction (*PERR*) criteria:

$$\mathrm{PER} = \frac{S+D+I}{N} \times 100\,[\%],$$
$$\mathrm{PERR} = \frac{\mathrm{PER}_{base} - \mathrm{PER}_{imp}}{\mathrm{PER}_{base}} \times 100\,[\%], \tag{1}$$

where *S, D, I*, and *N* represent the number of substituted, deleted, inserted, and total number of phones, respectively. PER_{imp} represents the improved error rate for the particular modelling technique and PER_{base} the base error rate against which the relative improvement was computed. In addition, the recognized transcription was remapped into three phonetic classes: voiced consonants, unvoiced consonants and vowels, and the phone error rate contribution for a particular phonetic class (PERC_{cl}) was computed using the following formula:

$$\mathrm{PERC}_{cl} = \frac{\mathrm{PER}_{cl}}{\mathrm{PER}_{all}},\,[\%] \tag{2}$$

The purpose was to evaluate the effect of compression on each phonetic class separately.

The initial study of the behavior of PLP and MFCC features for MP3 speech recognition was performed with all previously discussed AM refinement techniques but without any non-standard modifications to the feature extraction process. This analysis served as the benchmark for the subsequent modification in the form of additional dithering and its potential contribution.

Table 1 presents results for the PLP-based system. Significant differences in absolute PER were observed between compressed and uncompressed data for initial baseline AMs, but implementation of each subsequent modelling technique decreased the ΔPER. The studied bit-rates were selected to have a linear trend, but the achieved results marked the 20 kbps bit-rate as a breakpoint after which the error started to rise exponentially. This conclusion held true for all levels of AM development.

Table 1. PER [%] for PLP-based system for progressively refined AM

Data	Baseline	LDA	SAT	SGMM	MMI
Raw	17.9	16.1	12.3	10.2	7.2
128k	18.8	17.1	13.5	10.8	7.9
32k	19.3	17.3	13.2	11.0	8.2
28k	19.6	17.7	13.5	11.3	8.6
24k	20.7	18.7	14.1	11.9	9.3
20k	23.9	21.1	15.5	12.9	10.5
16k	36.2	30.4	18.6	15.5	13.3
12k	62.5	52.9	26.8	22.2	20.2

Another point of interest was the reduction of error as a function of the employed modelling technique. The fMLLR adaptation achieved the highest PERR for compressed speech in general, and its gain rose with decreasing bit-rate. These results indicate that the AM adaptation represents a crucial part of any system intended for MP3 recognition. On the other hand, the gain of discriminative training was the highest for raw data and decreased with decreasing bit-rate. This finding is consistent with the theoretical premise that discriminative training fits the AM on the training set but not necessarily on the testing set. In the case of our experiment, the MMI criteria optimized the AM for uncompressed signals and as the bit-rate decreased, so did the match between the testing and training signals. It should be noted, however, that the overall PERR computed for the baseline and final MMI models were in the 〈59 %, 67 %〉 range.

The same set of experiments for an MFCC-based system is summarized in Table 2. The system behaved similarly and displayed the same trends as far as the contribution of specific acoustic modelling techniques went. The PER displayed a tendency to rise rapidly after passing the 20 kbps breakpoint and AM adaptation proved to be crucial as it contributed the most to the overall PERR. The major difference was the overall increase of error rate for uncompressed signals. This finding leads to the conclusion that the MFCC features are not suitable for low bit-rate MP3 speech recognition.

Table 2. PER [%] for MFCC-based system for progressively refined AM

Data	Baseline	LDA	SAT	SGMM	MMI
Raw	17.8	15.9	12.3	10.3	7.3
128k	18.9	17.2	13.7	10.8	8.1
32k	20.9	17.9	13.6	11.3	8.7
28k	24.6	19.5	14.3	11.9	9.4
24k	33.9	25.9	16.4	13.5	11.4
20k	47.1	31.9	19.0	15.6	13.6
16k	62.2	49.4	25.8	20.4	18.7
12k	73.0	68.7	46.0	32.6	30.2

A more detailed study of the nature of error confirmed the theoretical assumptions about the compression distortions and their effect on particular phones outlined in Section 3. Figure 1 documents a decrease in PERC for voiced phones at the expense of unvoiced phones. PERC for unvoiced phones at the expense of voiced phones. While the reference PERC distribution was dominated by the vowels, the PERC for unvoiced constants steadily increased up to 34.2 %. Later experiments with partial contribution of each distortion showed that this negative effect occurred due to the combined presence of both the bandwidth limitation and spectral valleys.

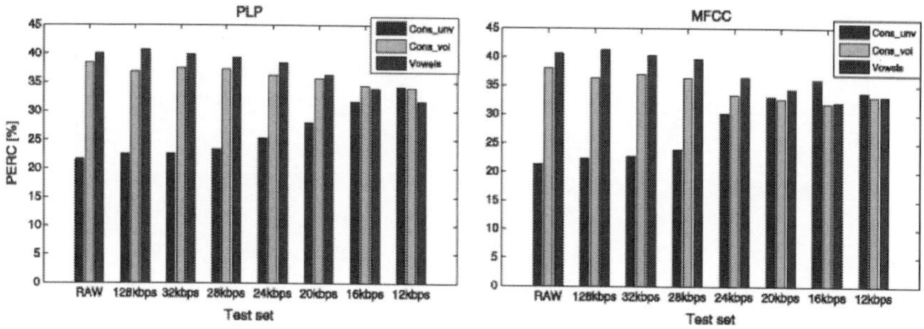

Figure 1. PERC [%] for PLP and MFCC features. The contribution of three phonetic groups: unvoiced consonants, voiced consonants, and vowels to the total phone error. The contribution of unvoiced consonants increased with the decreasing bit-rate

3.3 Results of additional dithering

All the above experiments used features without any further optimization. Since the previous runs showed the detrimental effect on higher frequency bands, it was important to investigate the proposed patch methods. The dithering value *R* was gradually increased by a factor of 2. Tables 3 and 4 present the best PER together with the optimal *R*.

Table 3. Results for dithered PLPs, the dithering value, and the corresponding error rate *R*/PER [%]

Bit-rate	Baseline	LDA	SAT	SGMM	MMI
128	2/18.7	2/17.0	4/13.2	4/10.7	4/ 7.9
32k	4/19.1	2/17.3	2/13.2	2/11.0	2/8.2
28k	4/19.4	4/17.7	4/13.5	4/11.2	4/8.6
24k	2/20.8	2/18.8	4/14.2	2/11.8	2/9.3
20k	4/23.5	2/21.0	4/15.6	2/12.9	2/10.5
16k	16/32.0	8/27.9	8/18.5	4/15.4	2/13.4
12k	32/44.7	32/39.4	16/24.9	8/21.2	4/19.8

Table 4. Results for dithered MFCCs, the dithering value, and the corresponding error rate *R*/PER [%]

Data	Baseline	LDA	SAT	SGMM	MMI
128k	2/18.9	2/17.1	4/13.1	4/10.7	4/8.0
32k	8/20.1	4/17.8	4/13.4	4/11.1	2/8.6
28k	8/21.6	4/18.6	8/13.9	8/11.5	8/9.1
24k	16/30.1	2/26.2	8/15.8	8/13.1	8/10.9
20k	16/34.7	8/30.0	8/17.7	8/14.8	8/12.8
16k	32/41.2	32/35.9	16/21.3	8/17.9	8/16.4
12k	64/49.9	64/45.5	16/29.0	16/24.2	16/23.0

Additional dithering for PLP features, Table 3, yielded consistent improvement for the lowest 12-kbps rate and some improvement for the 16-kbps rate. Its application was particularly useful for baseline and LDA models, but the reduction for more advanced acoustic modelling techniques was only marginal and higher bit-rates were mainly unaffected by the method. In cases where the dithering value was too high, the additional noise degraded the features further, which resulted in worse PER than for the undithered system. The process of estimating the *R* value included several iterations of feature extraction and decoding and thus consumed a lot of time and resources. When all of these factors were taken into consideration, we came to the conclusion that the usage of additional dithering cannot be advised for PLP features.

The next main point of interest was to investigate whether the dithered MFCCs can match the PLPs. These experiments showed more convincing results as positive PERR was obtained for all bit-rates and levels of AM refinement, as summarized in Table 4. The generally observed trend was that the lower bit-rates gained more from the additional dithering than the higher bit-rates. It should be noted, however, that MFCCs still did not manage to outperform the PLPs. The error rates were somewhere between the original MFCCs and PLPs.

Since it was confirmed that additional dithering can improve the recognition with MFCCs, the last analysis was focused on phonetic composition of the error. The initial hypothesis was that the addition of relatively weak noise would reconstruct the spectra of unvoiced phones, but the results showed that the reduction was spread evenly among phonetic classes or slightly towards the voiced phones. Figure 2 compares PERC for dithered and undithered MFCCs at 16/12 kbps as these were the only bit-rates which displayed a statistically relevant improvement. The values for the 16-kbps test sets with and without dithering were basically the same, which indicated a uniform contribution. The error for the dithered 12 kbps was dominated by the unvoiced phones, but the PERCs for the original MFCCs were even, which means that voiced phones gained more than the unvoiced phones. Based on these results, it can be speculated that the key principle of the method lies more in enhancing the feature's suitability for statistical modelling and less in actual reconstruction of their spectral characteristics.

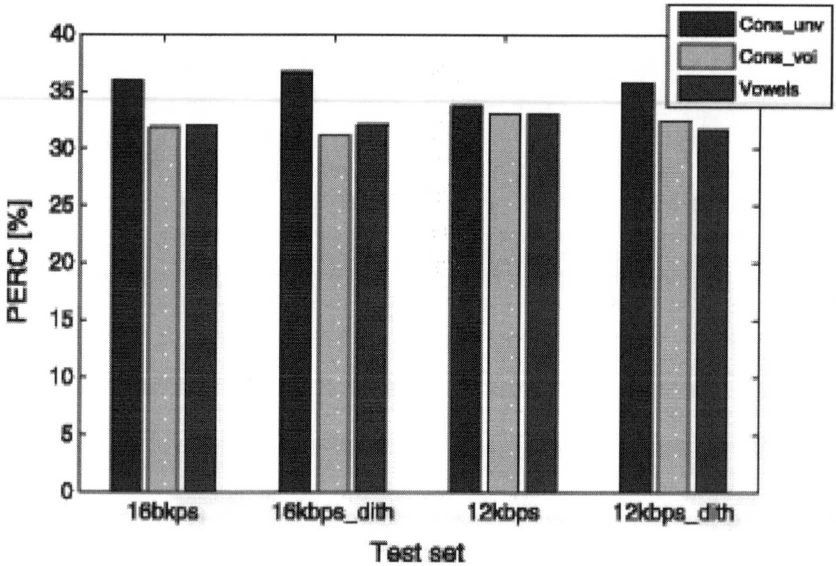

Figure 2. The comparison of PERC [%] for classic and additionally dithered MFCC features. The analysis showed little difference in PERC distribution between classic and modified features. The contribution of additional dithering was constant for every class

3.4 Results of large vocabulary continuous speech recognition

Since MP3 speech recognition is generally intended for applications such as off-line transcription of recorded speech or indexing of audio archives, the following experiments were aimed at analyzing the described acoustic modelling techniques at the standard large vocabulary continuous speech recognition (LVCSR) task. The used AMs were trained in the described manner, the decoding graph was constructed from the bigram language model [16] of 340k vocabulary size created from the Czech National Corpus [17], and the test subset with an overall length of 1 h contained only signals with a full sentence structure. The results were evaluated by the standard word error rate (WER) metric.

Table 5 compares the LVCSR results for the baseline, adapted, and MMI trained AMs. Since the dithered PLPs displayed practically the same error rates as the undithered ones, they were not included in the next set of experiments. The observed trends of error rate corresponded to the conclusions drawn from the phoneme experiments. The error started to rise exponentially after passing the 20-kbps threshold, and the AM adaptation was the main source of error reduction for lower bit-rates. The advanced acoustic modelling techniques displayed a trend of increasing relative gains as the bit-rates decreased, a conclusion which was in contrast the previous results. This development occurred most likely due to the usage of word level LM in combination with progressively better AMs. The PLP features achieved better results than MFCC and marginally better than the dithered MFCC (dMFCC) features.

Previous experiments determined that lossy compression affects unvoiced phonemes more strongly than the voiced ones. In order to determine which compression degradation, bandwidth limitation, or spectral valleys is more detrimental to the overall performance, we decided to estimate their partial contributions separately. The uncompressed signals were filtered by a FIR low-pass filter at corresponding cutoff frequencies, which were estimated for each bit-rate from their spectrograms. The spectral valleys' contribution was estimated by replacing the suppressed spectral bins at higher frequencies of a coded signal with the spectral bins from an uncoded signal. The replaced bins were selected using the same cut-off frequencies. It should be noted however, that other compression artifacts (pre-echo, birdie, etc.) might have also degraded the lower bands and thus affected the obtained results. This approach allowed us to quantify the contributions as if they affected only the selected parts of the spectra.

Table 6 summarizes the results of recognition in the LVCSR task and gives an overview of the used cutoff frequencies of the filters. The results demonstrate that the spectral valleys degraded the speech more significantly on average, but that their contribution to the overall degradation was only marginal, with the exception of the 12-kbps bit-rate and MFCC features. The performance drop generally observed in the MP3 speech was the result of the non-linear combination of both distortions.

Table 5. WER [%] for LVCSR task with 340k bigram LM

Features	Raw	128k	32k	28k	24k	20k	16k	12k
PLP_base	23.74	24.5	24.5	25.57	25.98	28.19	38.79	68.57
PLP_adapt	18.19	19.45	19.4	19.59	19.84	20.67	23.56	33.19
PLP_MMI	14.25	14.43	14.55	14.54	15.21	16.15	18.57	25.23
MFCC_base	23.72	25.07	25.13	26.67	31.75	38.46	62.45	91.43
MFCC_adapt	18.44	19.06	19.11	19.92	20.7	22.51	28.11	44.82
MFCC_MMI	14.22	14.72	14.92	15.12	15.82	17.57	21.48	31.54
dMFCC_base	–	24.85	25.05	26.01	31.77	36.69	48.83	70.03
dMFCC_adapt	–	18.75	19.01	19.61	20.32	21.71	25.17	34.75
dMFCC_MMI	–	14.25	14.78	15.06	15.5	16.84	19.47	26.41

Table 6. The partial contribution of LP filtering and spectral valleys (SV) on LVCSR task, WER [%]

	Bit-rate	128k	32k	28k	24k	20k	16k	12k
	f_{cut}		7200 Hz			5800 Hz		5600 Hz
PLP	SV	14.16	14.28	14.45	14.22	14.43	14.96	16.68
	Low-pass		14.22			14.34		14.54
MFCC	SV	14.35	14.91	14.89	14.94	15.56	16.18	19.7
	Low-pass		14.15			14.45		14.86

4. CONCLUSIONS

This paper studied the current state-of-the-art acoustic modelling and a specific feature compensation technique in the task of MP3 speech recognition. More precisely, linear discriminant analysis in conjunction with acoustic model adaptation, subspace Gaussian mixture model framework, discriminative training, and additional dithering was described. The baseline system was trained on uncompressed data and tested on both the uncompressed and the compressed signals in the task of phoneme recognition and LVCSR.

The evaluation runs documented that the usage of PLP features and application of AM adaptation and discriminative training can reduce the error rate of the system. The MMI-trained AMs performed at 14.24 % on the reference test set, but the WER dropped to 18.57 % for 16 kbps and 25.23 % for 12 kbps rates. For comparison, the MFCC system performed at 14.22, 21.48, and 31.54 % WERs for the same subsets. Adapting the AMs to the specific speaker and bit-rate yielded the highest mean improvement out of all the analyzed modelling techniques and proved to be essential for recognition of compressed speech. Our preliminary experiments on usage of DNN-HMM displayed results which were slightly worse (by approximately 1 %) than GMM-HMM and, for this reason, were not presented in article.

The phoneme-level recognition confirmed the theoretical hypothesis that the MP3 compression affects the unvoiced phonemes more significantly than voiced ones phonemes. The contribution of the unvoiced phonemes to the total phone error rose from 21.6 % for the reference test set to 34.2 % for the 12-kbps set. A more detailed study of bandwidth limitation and spectral valleys showed that the observed increase of the error rate occurred as a result of the non-linear combination of both distortions.

While the observed results justified the usage of additional dithering for the MFCC features, the error rates for dithered MFCCs were still slightly higher than those for PLPs. However, the main problem of this approach was the need to manually tune the dithering value to achieve the best results. The results of detailed phoneme accuracy showed that the technique was not able to compensate for the loss of information due to the low-pass filtering and spectral valleys phenomena.

ACKNOWLEDGEMENTS

Research described in the paper was supported by internal CTU Grant SGS14/191/OHK3/3T/13 "Advanced Algorithms of Digital Signal Processing and their Applications".

COMPETING INTERESTS

The authors declare that they have no competing interests.

REFERENCES

1. C-M Liu, H-W Hsu, W-C Lee, Compression artifacts in perceptual audio coding. IEEE Trans. Audio Speech Lang. Process. 16(4), 681–695 (2008).

2. RJJH Van Son, A study of pitch, formant, and spectral estimation errors introduced by three lossy speech compression algorithms. Acta Acustica United Acustica. 91, 771–778 (2005).

3. C Barras, L Lamel, J Gauvain, in *Acoustics, Speech, and Signal Processing. Proceedings of 2001 IEEE International Conference on*, 1. Automatic transcription of compressed broadcast audio (Salt Lake City, USA, 2001), pp. 265–268.

4. L Besacier, C Bergamini, D Vaufreydaz, E Castelli, in *Multimedia Signal Processing. Proceedings of 2001 IEEE Fourth Workshop on*. The effect of speech and audio compression on speech recognition performance (Cannes, France, 2001), pp. 301–306.

5. PS Ng, I Sanches, in *Proceedings of 2004 Conference Speech and Computer, SPEECOM*. The influence of audio compression on speech recognition systems (St. Petersburg, Russia, September 2004).

6. P Pollak, M Borsky, Communications in Computer and Information Science, in *E-Business and Telecommunications*. Small and large vocabulary speech recognition of MP3 data under real-word conditions: experimental study, vol.314 (SpringerBerlin, 2012), pp. 409–419.

7. J Nouza, P Cerva, J Silovsky, in *Acoustics, Speech and Signal Processing. Proceedings of 2013 IEEE International Conference on*. Adding controlled amount of noise to improve recognition of compressed and spectrally distorted speech (Vancouver, Canada, May 2013), pp. 8046–8050.

8. H Abbasian, B Nasersharif, A Akbari, M Rahmani, MS Moin, in *Communications, Control and Signal Processing. 3rd International Symposium on*. Optimized linear discriminant analysis for extracting robust speech features, (March 2008), pp. 819–824.

9. S Tamura, S Hayamizu, in *Signal Information Processing Association Annual Summit and Conference. 2012 Asia-Pacific*. Multi-stream acoustic model adaptation for noisy speech recognition (Hollywood, USA, December 2012), pp. 1–4.

10. U Remes, KJ Palomäki, M Kurimo, in *EUSIPCO. Proceedings of 16th European Signal Processing Conference*. Missing feature reconstruction And Acoustic Model Adaptation Combined For large vocabulary continuous speech recognition (Lausanne, Switzerland, 2008).

11. D Yu, L Deng, Y Gong, A Acero, in *Proceedings of the Interspeech*. Discriminative training of variable-parameter HMMs for noise robust speech recognition (International Speech Communication Association Brisbane, Australia, September 2008).

12. J Du, P Liu, F Soong, J-L Zhou, R-H Wang, Lecture Notes in Computer Science, in *Chinese Spoken Language Processing*. Noisy speech recognition performance of discriminative HMMs vol.4274 (SpringerBerlin, 2006), pp. 358–369.

13. D Povey, A Ghoshal, G Boulianne, L Burget, O Glembek, N Goel, M Hannemann, P Motlicek, Y Qian, P Schwarz, J Silovsky, G Stemmer, K Vesely, in *IEEE 2011 Workshop on Automatic Speech Recognition and Understanding*. The Kaldi speech recognition toolkit (IEEE Signal Processing SocietyHilton Waikoloa Village, Big Island, Hawaii, US, 2011).

14. R Hegemann, A Leidinger, R Brito, LAME (2011). http://lame.sourceforge.net

15. P Fousek, P Mizera, P Pollak, CtuCopy feature extraction tool (2014). http://noel.feld.cvut.cz/speechlab

16. V Prochazka, P Pollak, J Zdansky, J Nouza, Performance of Czech speech recognition with language models created from public resources. Radioengineering. 20, 1002–1008 (2011).

17. Ústav Českého národního korpusu FF UK Praha, Český národní korpus - SYN2006PUB (2006). http://www.korpus.cz

Chapter 4

Evaluation of Random-Projection-Based Feature Combination on Dysarthric Speech Recognition

Toshiya Yoshioka, Tetsuya Takiguchi, Yasuo Ariki

Graduate School of System Informatics, Kobe University, 1-1 Rokkodai, Nada, Kobe, 6578501, Japan

ABSTRACT

We investigated the speech recognition of persons with articulation disorders resulting from cerebral palsy. The articulation of their first speech tends to become unstable due to strain on speech-related muscles, and that causes degradation of speech recognition. In this paper, we propose a feature extraction method based on RP (Random Projection) for dysarthric speech recognition. Random projection has been suggested as a means of space mapping, where the original data are projected onto a space using a random matrix. It represents a computationally simple method that approximately preserves the Euclidean distance of any two points through the projection. Moreover, as we are able to produce various random matrices, there may be some possibility of finding a random matrix that gives better speech recognition accuracy among these random matrices. To obtain an optimal result from many random matrices, a vote-based combination is introduced in this paper. ROVER combination is applied to the recognition results obtained from the ASR (Automatic Speech

Recognition) systems created from each RP-based feature. Its effectiveness is confirmed by word recognition experiments.

KEYWORDS

Articulation Disorders, Speech Recognition, Random Projection, ROVER

1. INTRODUCTION

Recently, the importance of information technology in the welfare-related fields has increased. For example, sign language recognition using image recognition technology [1][2][3], text-reading systems from natural scene images [4][5][6], and the design of wearable speech synthesizers for voice disorders[7][8] have been studied.

There are 34,000 people with speech impediments associated with articulation disorders in Japan alone, and it is hoped that speech recognition systems will one day be able to recognize their voices. One of the causes of speech impediments is cerebral palsy. Cerebral palsy results from damage to the central nervous system, and the damage causes movement disorders. Three general times are given for the onset of the disorder: before birth, at the time of delivery, and after birth. Cerebral palsy is classified as follows: 1) spastic type 2) athetoid type 3) ataxic type 4) atonic type 5) rigid type, and a mixture of types[9].

In this paper, we focused on persons with articulation disorders resulting from the athetoid type of cerebral palsy. Athetoid symptoms develop in about 10-15% of cerebral palsy sufferers. In the case of a person with this type of articulation disorder, the first movements are sometimes more unstable than usual. That means, in the case of speaking-related movements, the first utterance is often unstable or unclear due to the athetoid symptoms, and that causes degradation of speech recognition. Therefore, we recorded speech data for persons with articulation disorders who uttered each of the words several times, and investigated the influence of the unstable speaking style caused by the athetoid symptoms.

The goal of front-end speech processing in ASR is to obtain a projection of the speech signal to a compact parameter space where the information related to speech content can be extracted. In current speech recognition technology, MFCC (Mel-Frequency Cepstrum Coefficient) is being widely used. The feature is uniquely derived from the mel-scale filter-bank output by DCT (Discrete Cosine Transform). The low-order MFCCs account for the slowly changing spectral envelope, while the high-order ones describe the fast variations of the spectrum. Therefore, a large number of MFCCs is not used for speech recognition because we are only interested in the spectral envelope, not in the fine structure. In[10], we proposed robust feature extraction based on PCA (Principal Component Analysis) with more stable utterance data instead of DCT in a dysarthric speech recognition task. Also,[11] used MAF (multiple acoustic frames) as an acoustic dynamic feature to improve the recognition rate of a person with articulation disorders, especially in speech recognition using dynamic features only. These methods improved the recognition accuracy, but the performance for articulation disorders was not sufficient when compared to that of persons with no disability.

Random projection has been suggested as a means of space mapping, where a projection matrix is composed of the columns defined by the random values chosen from a probability distribution. In addition, the Euclidean distance of any two points is approximately preserved through the projection. Therefore, random projection has also been suggested as a means of dimensionality reduction[12]. In contrast to conventional techniques such as PCA, which find a subspace by optimizing certain criteria, random projection does not use such criteria; therefore, it is data independent. Moreover, it represents a computationally simple and efficient method that preserves the structure of the data without introducing significant distortion[13]. Goel et al[13] have reported that random projection has been applied to various types of problems, including information retrieval (e.g.,[14]), image processing (e.g.,[15][16]), machine learning (e.g.,[17][18][19]), and so on. Although it is based on a simple idea, random projection has demonstrated good performance in a number of applications, yielding results comparable to conventional dimensionality reduction techniques, such as PCA.

In this paper, we investigate the feasibility of random projection for speech feature transformation in order to improve the recognition rate of persons with articulation disorders. In our proposed method, original speech features (MFCCs) are transformed using various random matrices. Then, we use the same number of dimensions for the projected space as that of the original space. There may be some possibility of finding a random matrix that gives better speech recognition accuracy among random matrices, since we are able to produce various RP-based features (using various random matrices). Therefore, a vote-based combination method is introduced in order to obtain an optimal result from many (infinite) random matrices, where ROVER combination[20] is applied to the results from the ASR systems created from each RP-based feature.

The rest of the paper is organized as follows. Section 2 describes a feature projection method using random orthogonal matrices. In Section 3, a vote-based combination method is explained. Results and discussion for the experiments on a dysarthric speech recognition task are given in Section 4. Section 5, concludes the paper with a summary of our proposed method, contribution, and future work.

2. RANDOM ORTHOGONAL PROJECTION

This section presents a feature projection method using random orthogonal matrices. The main idea of random projection arises from the Johnson-Lindenstrauss lemma [21]; namely, if original data are projected onto a randomly selected subspace using a random matrix, then the distances between the data are approximately preserved.

Random projection is a simple yet powerful technique, and it has another benefit. Dasgupta[17] has reported that even if distributions of original data are highly skewed (have ellipsoidal contours of high eccentricity); their transformed counterparts will be more spherical.

First, we choose an $n-$ dimensional random vector, p, and let $\mathrm{P}^{(i)}$ be the i - th $n \times d$ matrix whose columns are vectors, $\mathrm{p}_1^{(i)}, \mathrm{p}_2^{(i)}, \cdots, \mathrm{p}_d^{(i)}$. Then, an original $n-$ dimensional vector, x, is projected onto a $d-$dimensional subspace using the i - th random matrix, $\mathrm{P}^{(i)}$, where we compute a $d-$ dimensional vector, x', whose coordinates are the inner products, $\mathrm{x}_1' = \mathbf{P}_1^{(i)} \cdot \mathrm{x}, \cdots, \mathrm{x}_d' = \mathbf{P}_d^{(i)} \cdot \mathrm{x}$.

$$\mathrm{x}' = \mathbf{P}^{(i)^{\mathrm{T}}} \mathrm{x} \tag{1}$$

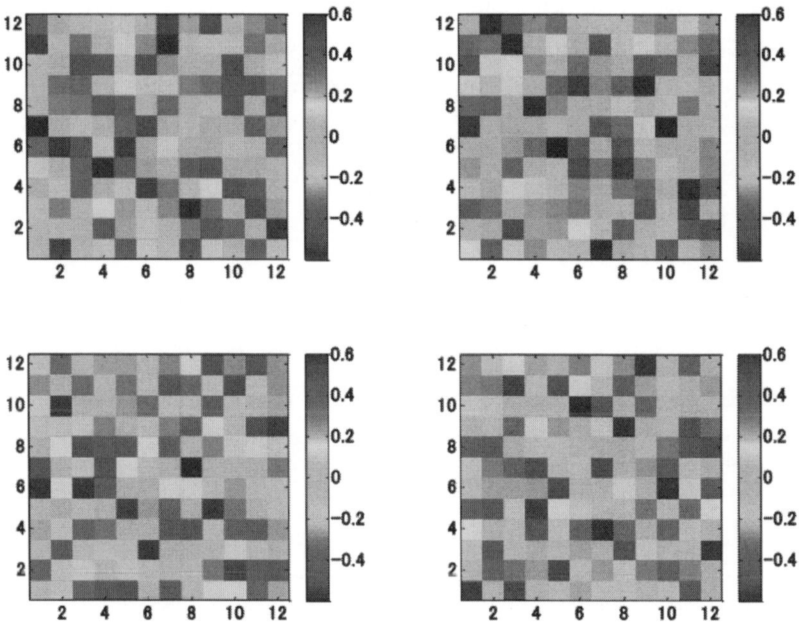

Figure 1. Examples of random matrices 12 dim. (12 12)

In this paper, we investigate the feasibility of random projection for speech feature transformation. As described above, a random projection from n dimensions to d (=n) dimensions is represented by an $n \times d$ matrix, $\mathbf{P}$. It has been shown that if the random matrix, $\mathbf{P}$, is chosen from the standard normal distribution, with mean 0 and variance 1, referred to as $N(0,1)$. Then, the projection preserves the structure of the data[21]. In this paper we use $N(0,1)$ for the distribution of the coordinates. The random matrix, $\mathbf{P}$, can be calculated using the following algorithm [13][17].

- Choose each entry of the matrix from an independent and identifiably distributed $N(0,1)$ value.
- Make the orthogonal matrix using the Gram Schmidt algorithm, and then normalize it to unit length.

Orthogonality is effective for feature extraction because the HMMs used in speech recognition experiments consist of diagonal covariance matrices. Fig. 1 shows examples of random matrices from $N(0,1)$.

3. VOTE-BASED COMBINATION

As described in Section 2, we can make many (infinite) random matrices from $N(0,1)$ (Fig. 2). Since there may be some possibility of finding a random matrix that gives better performance, we will have to select the optimal matrix or the optimal recognition result from them. To obtain the optimal result, a majority vote-based combination is introduced in this paper, where ROVER combination is applied to the results from the ASR systems created from each RP-based feature.

Fig. 3 shows an overview of the vote-based combination method. First, random matrices $\mathbf{P}^{(l)}$ ($l = 1, \cdots, L$) are chosen from the standard normal distribution, with mean 0 and variance 1. Original speech features (MFCCs) are projected using each random matrix. An acoustic model corresponding to each random matrix is also trained. For the test utterance, using each acoustic model, an ASR system outputs the best scoring word by itself. To obtain a single hypothesis from among all the results for random projection, voting is performed by counting the number of occurrences of the best word for each RP-based feature.

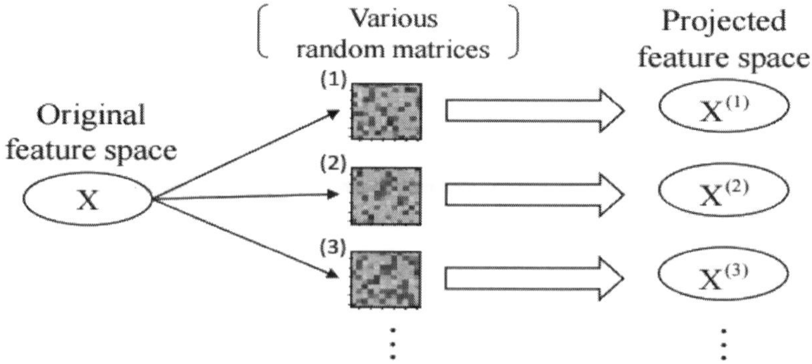

Figure 2. Random projection on the feature domain. An original feature is transformed to various features using various random matrices. (Eq. 1

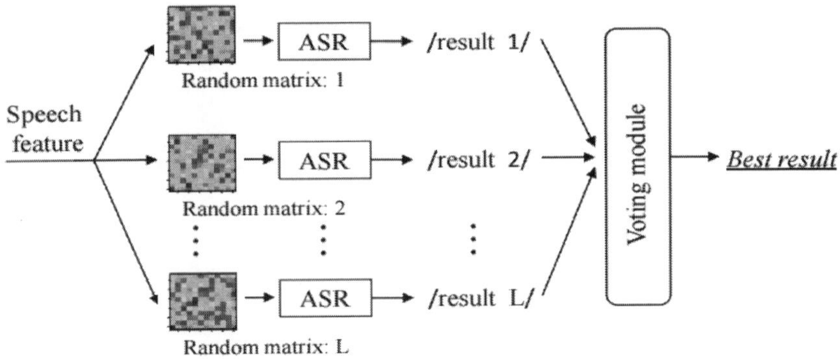

Figure 3. Overview of the vote-based combination

For example, in the case of l = 20, 20 kinds of new feature vectors are calculated using 20 kinds of random matrices. Then, we train the 20 kinds of acoustic models using 20 kinds of new feature vectors. In the test process, 20 kinds of recognition results are obtained using 20 kinds of acoustic models. To obtain a single hypothesis from among 20 kinds of recognition results, voting is performed.

4. EVALUATION

4.1. Experimental Conditions

The proposed method was evaluated on a word recognition task for three males with articulation disorders (Speaker A, B, C). For the conducted experiments, we recorded 210 words included in the ATR Japanese speech database for each speaker. Each of the 210 words was repeated five times (Fig. 4). The speech signal was sampled at 16 kHz and windowed with a 25-msec Hamming window every 10 msec. Fig. 5 shows an example of a spectrogram spoken by a person with an articulation disorder. Fig. 6 shows a spectrogram spoken by a physically unimpaired person doing the same task.

The recognition results for a speaker-independent model are shown in Fig. 7, where the speech data uttered by unimpaired persons were used. As can be seen in Fig. 7, the recognition rate of a physically unimpaired person was around 90%. However, the result of a person with an articulation disorder (speaker A) was only 3.5%. It is clear that the speaking style of a person with an articulation disorder differs considerably from that of physically unimpaired persons.

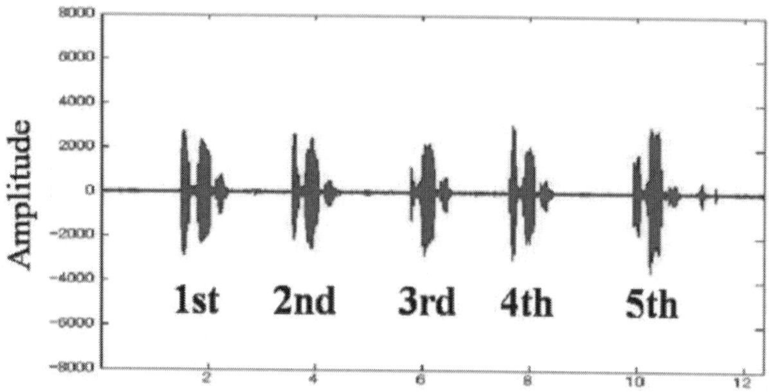

Figure 4. Example of recorded speech data

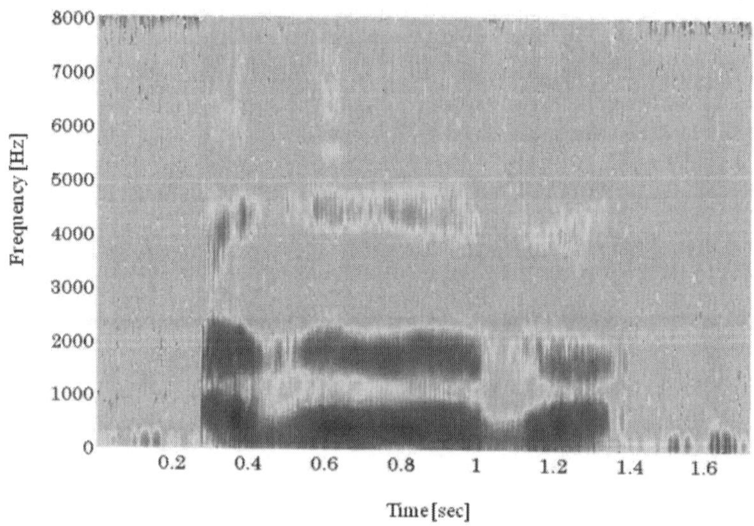

Figure 5. Example of a spectrogram from a person with an articulation disorder. (//a k e g a t a)

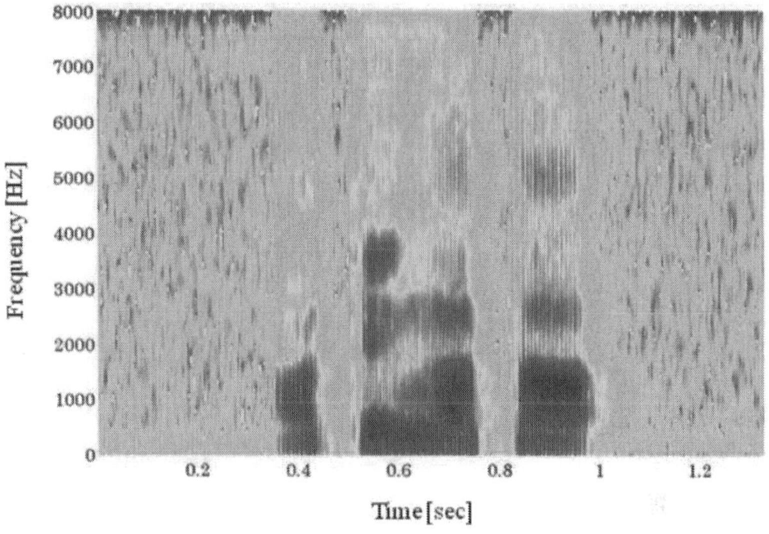

Figure 6. Example of a spectrogram from a physically unimpaired person. (//a k e g a t a)

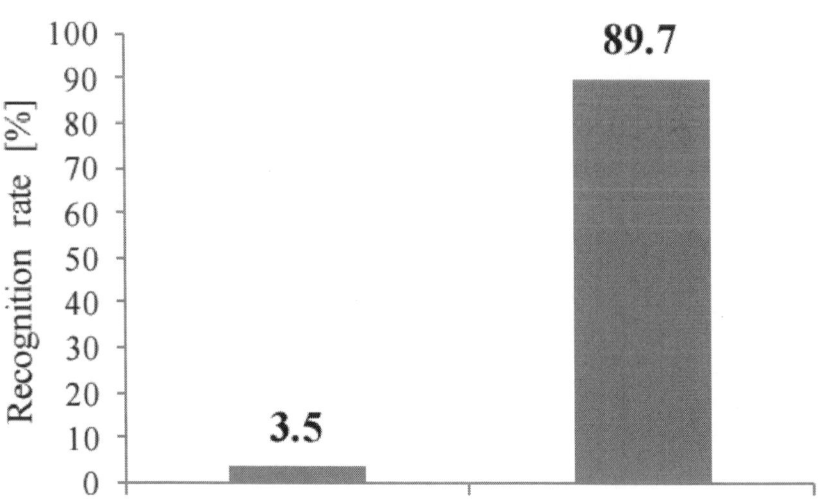

Figure 7. Recognition results[%] for the speaker-independent model using training data uttered by unimpaired persons

Also, Fig. 7 shows that it was difficult to recognize utterances of articulation disorders using an acoustic model trained by utterances of physically unimpaired persons. Therefore, in this paper, we trained the acoustic model using the utterances of a person with an articulation disorder. The acoustic model consists of a HMM set with 54 context-independent phonemes and 8 mixture components for each state. Each HMM has three states and three self-loops.

4.2. Experiment 1

In Experiment 1, recognition results were obtained for each utterance using speaker-dependent models. Each system was trained using 24-dimensional feature vectors consisting of 12-dimensional MFCC parameters, along with their delta parameters. When we recognized the 1st utterance, the 2nd through 5th utterances were used for training. We iterated this process for each utterance.

Table 1. Recognition results[%] for each utterance of speaker (A)

1st	2nd	3rd	4th	5th
75.7	86.7	92.9	90.5	88.6

Table 1 shows the results obtained in Experiment 1 for speaker (A). As can be seen in Table 1, the recognition rate for the 1st utterance was 75.7%. It was significantly lower than other utterances. It is considered that the speaker experiences a more strained state during the first utterance compared to subsequent utterances because the first utterance is the first intentional movement. Therefore, athetoid symptoms occur and articulation becomes difficult. It is believed that this difficulty causes fluctuations in speaking style and degradation of the recognition rates.

Tables 2 and 3 show the recognition results for each utterance for speaker (B) and (C), respectively. As can be seen in Tables 2 and 3, a decrease in recognition rate for the first utterance due to fluctuations in speaking style was confirmed.

Table 2. Recognition results[%] for each utterance of speaker (B)

1st	2nd	3rd	4th	5th
85.7	91.9	91.4	93.3	95.2

Table 3. Recognition results[%] for each utterance of speaker (C)

1st	2nd	3rd	4th	5th
94.3	99.5	97.6	97.6	95.7

4.3. Experiment 2

The aim of Experiment 2 is to evaluate the improvement introduced by the use of a RP-based feature projection method for the unstable 1st utterance. In the experiments, the following RP-based features were evaluated. Random projection is applied to MFCC at the t-th frame, $x(t) \in R^{12}$, and the new feature, $y(t) \in R^{12}$, is obtained:

$$y(t) = P^{(i)T} x(t) \tag{2}$$

Then, the new feature also has the delta parameter of original featur, $x(t)$. The final system feature dimensionality is 24 ($MFCC[12] \rightarrow RP[12] + \Delta MFCC[12]$).

We investigated the performance of random projections for various random matrices (*I* = 20, 40, 60, 80, and 100) sampled from $N(0,1)$. Tables 4 ~ 6 show the recognition rate versus the number of random matrices for each speaker. The results of "RP w/o combination" show the maximums, means, and minimums obtained from each random projection without ROVER-based combination.

Table 4 shows the performance results versus the number of random matrices for speaker (A). As can be seen in Table 4, the results for RP-based feature indicate that the vote-based random-projection combination improved the recognition rate from 75.7% to 81.9% using the combination of 60 random matrices, and even the means of random projections without combination for some random matrices was better than the recognition rate of the original features. Also, even if the number of random matrices is changed, we do not see that subsequent performance varies in our experiments.

Table 4. Word recognition rate[%] for the 1st utterances of speaker (A) using the proposed method in various random matrices. (The recognition rate for the original features is 75.7%)

Number of random matrices	RP combination based on ROVER	RP w/o combination		
		Max.	Mean	Min.
20	80.5	80	76.8	73.8
40	81.4	80	76.8	72.9
60	**81.9**	80.5	77	72.9
80	81.4	80.5	76.9	72.9
100	81	80.5	76.8	72.9

Table 5. Word recognition rate[%] for the 1st utterances of speaker (B) using the proposed method in various random matrices. (The recognition rate for the original features is 85.7%)

Number of random matrices	RP combination based on ROVER	RP w/o combination		
		Max.	Mean	Min.
20	90	92.9	86.8	84.3
40	89.1	92.9	87.1	84.3
60	**90.5**	92.9	87.1	84.3
80	**90.5**	92.9	87.1	84.3
100	90	92.9	87.1	84.3

Tables 5 and 6 show the performance of the proposed method for speaker (B) and (C), respectively. As can be seen in Table 5, the recognition rate of 90.5% was obtained using the combination of 60 or 80 random matrices. Also, the results in Table 6 were maintained above 95%, showing the effectiveness of our proposed method for each person with articulation disorders who participated in our experiment.

Table 6. Word recognition rate[%] for the 1st utterances of speaker (C) using the proposed method in various random matrices. (The recognition rate for the original features is 94.3%)

Number of random matrices	RP combination based on ROVER	RP w/o combination		
		Max.	Mean	Min.
20	96.2	97.1	95.3	93.8
40	95.7	97.1	95.3	93.8
60	**96.7**	97.6	95.5	93.8
80	**96.7**	97.6	95.5	93.3
100	95.7	97.6	95.5	93.3

4.4. Experiment 3

In order to show the superiority of the RP-based feature projection method, in Experiment 3, we compared the proposed method and the PCA-based feature projection method.

For Experiment 3, PCA was applied to 12-dimensional MFCC, and the new feature also had the delta coefficient of the MFCC features. Then, we computed the eigenvector matrix using the 2nd through 5th utterances (the more stable utterances) for each speaker. A comparison between the PCA-based feature projection method and the results obtained by our proposed method using the combination of 60 random matrices are shown in Fig. 8.

We can see that the combination of random projection and ROVER outperforms both the baseline method (MFCCs) and the PCA-based feature extraction method. This result gives the evidence of the improvement introduced by the speech feature extraction based on random projection and the use of ROVER to obtain an optimal result. One of the possible reasons the random projection improves the recognition rates may be that if distributions of original data are skewed (have ellipsoidal contours of high eccentricity), their transformed counterparts will become more spherical [17]. However, there were 'bad' projections that cause degradation of speech recognition accuracy compared with the recognition of original features (Tables 4 ~ 6). Therefore, more research will be needed to investigate the effectiveness of the random projection method for speech features.

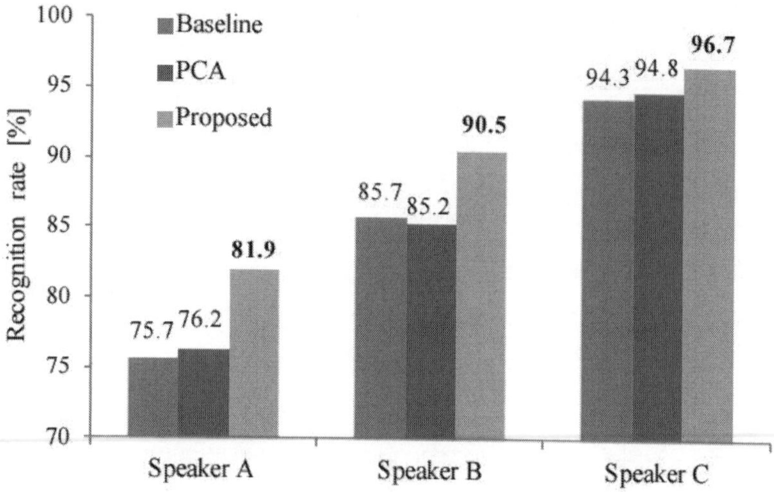

Figure 8. Comparison between the PCA-based feature projection method and the results obtained by our proposed method for the 1st utterance

5. CONCLUSIONS

As a result of this work, a method for recognizing dysarthric speech using RP-based features has been developed. The proposed method transforms the conventional speech features such as MFCC using various random matrices. It also introduces the vote-based combination method to obtain an optimal result from the ASR systems created from each RP-based feature. Word recognition experiments were conducted to evaluate the proposed method for three males with articulation disorders. The results of the experiments showed that all the recognition rates of the proposed method outperformed the baseline rate (using MFCCs).

As future work, we will continue to investigate how to select the optimal basis vector via the use of such random matrices.

ACKNOWLEDGEMENTS

This research was supported in part by MIC SCOPE.

REFERENCES

1. J. Lin, W. Ying, and T. S. Huang, "Capturing human hand motion in image sequences," IEEE Motion and Video Computing Workshop, pp. 99-104, 2002.

2. T. Starner, J. Weaver, and A. Pentland, "Real-Time American Sign Language Recognition Using Desk and Wearable Computer Based Video," IEEE Transactions on Pattern Analysis and Machine Intelligence, 20(12), pp. 1371-1375, 1998.

3. G. Fang, W. Gao and D. Zhao, "Large vocabulary sign language recognition based on hierarchical decision trees," Proceedings of the 5th international conference on Multimodal interfaces, pp. 125-131, 2003.

4. N. Ezaki, M. Bulacu and L. Schomaker, "Text Detection from Natural Scene Images: Towards a System for Visually Impaired Persons," ICPR 2004, pp. 683-686, 2004.

5. V. Wu, R. Manmatha and E. M. Riseman, "Textfinder: an automatic system to detect and recognize text images," IEEE Transactions on Pattern Analysis and Machine Inteligence, 21(11), pp. 1224-1229, 1999.

6. M. K. Bashar, T. Matsumoto, Y. Takeuchi, H. Kudo, and N. Ohnishi, "Unsupervised Texture Segmentation viaWavelet-based Locally Orderless Images (WLOIs) and SOM," 6th IASTED International Conference COMPUTER GRAPHICS AND IMAGING, 2003.

7. T. Ohsuga, Y. Horiuchi, and A. Ichikawa, "Estimating Syntactic Structure from Prosody in Japanese Speech," IEICE Transactions on Information and Systems, 86(3), pp. 558-564, 2003.

8. K. Nakamura, T. Toda, H. Saruwatari, and K. Shikano, "Speaking Aid System for Total Laryngectomees Using Voice Conversion of Body Transmitted Artificial Speech," INTERSPEECH, pp. 1395-1398, 2006.

9. S.T. Canale, and W.C. Campbell, “Campbell's Operative Orthopaedics,” Mosby-Year Book, 2002.

10. H. Matsumasa, T. Takiguchi, Y. Ariki, I. Li, and T. Nakabayashi, “PCA-Based Feature Extraction for Fluctuation in Speaking Style of Articulation Disorders,” INTERSPEECH 2007, pp. 1565-1568, 2007.

11. C. Miyamoto, Y. Komai, T. Takiguchi, Y. Ariki, and I. Li, “Multimodal Speech Recognition of a Person with Articulation Disorders Using AAM and MAF,” 2010 IEEE International Workshop on Multimedia Signal Processing, pp. 517-520, 2010.

12. Ella Bingham, and Heikki Mannila, “Random projection in dimensionality reduction: applications to image and text data,” KDD’01 Proceedings of the seventh ACM SIGKDD international conference on Knowledge discovery and data mining, pp. 245-250, 2001.

13. N. Goel, G. Bebis, and A. Nefian, “Face Recognition Experiments with Random Projection,” SPIE, vol. 5779, pp. 426-437, 2005.

14. P. Thaper, S. Guha, and N. Koudas, “DynamicMultidimensional Histograms,” ACM SIGMOD, pp. 428-439, 2002.

15. L. Liu, P. Fieguth, G. Kuang, and H. Zha, “Sorted Random Projections for robust texture classification,” IEEE International Conference on Computer Vision, pp. 391-398, 2011.

16. H. T. Ho, and R. Chellappa, “Automatic head pose estimation using randomly projected dense SIFT descriptors,” IEEE International Conference on Image Processing, pp. 153-156, 2012.

17. S. Dasgupta, “Experiments with random projection,” UAI, pp. 143-151, 2000.

18. X. Z. Fern, and C. E. Brodley, “Random Projection for High Dimensional Data Clustering: A Cluster Ensemble Approach,” the 20th Int. Conf. on Machine Learning, pp. 186-193, 2003.

19. S. Lee, and A. Nedic, “Distributed Random Projection Algorithm for Convex Optimization,” IEEE Journal of Selected Topics in Signal Processing, Vol. 7, No. 2, pp. 221-229, 2013.

20. J. G. Fiscus, "A post-processing system to yield reduced word error rates: Recogniser output voting error reduction (ROVER)," IEEE ASRU, pp. 347-352, 1997.

21. R. I. Arriaga, and S. Vempala, "An algorithmic theory of learning: robust concepts and random projection," IEEE Symposium on Foundations of Computer Science, pp. 616-623, 1999.

Chapter 5

Improvement on Sound Quality of the Body Conducted Speech from Optical Fiber Bragg Grating Microphone

Masashi Nakayama[1, 2], Shunsuke Ishimitsu[3] and Seiji Nakagawa[2]

[1] Kagawa National College of Technology, Japan

[2] National Institute of Advanced Industrial Science and Technology (AIST), Japan

[3] Hiroshima City University, Japan

1. INTRODUCTION

Speech communication can be impaired by the wide range of noise conditions present in air. Researchers in the field of speech applications have been investigating how to improve the performances of signal extraction and its recognition in the conditions. However, it is not yet possible to measure clear speech in environments where there are low Signal-to-Noise Ratios (SNR) of about 0 dB or less (H. Hirsch and D. Pearce, 2000). Standard rate scales, such as CENSREC(N. Kitaoka et al., 2006) and AURORA (H. Hirsch and D. Pearce,2000), are typically discussed for evaluating performances of speech recognition in noisy environments and have shown thatspeech recognition rates are approximately 50–80% whenunder the influence of noise, demonstrating the difficulty ofachieving high percentages.With these backgrounds, many signal extraction and retrieval methods have been proposed in previous research. There is one of approaches in signal extractions, body-conducted speech (BCS) which is little influence from noise in air however it does not measure 2 kHz above in frequency characteristics. However, these need normal speech or parameters measured simultaneously with body-conducted speech. Because these parameters are not measured in noisy environments, the authors have been investigating the use of body-conducted speech which is generally called bone-conducted speech, where the signal is also conducted through the skin and bone in a human body (S. Ishimitsu, 2008) (M. Nakayama et al., 2011). Conventional retrieval methods for sound quality of body-conducted speech are the Modulation Transfer Function (MTF), Linear Predictive Coefficients (LPC), direct filtering and the use of a throat microphone (T. Tamiya, and T. Shimamura, 2006) (T. T. Vu et al., 2006) (Z. Liu et al., 2004) (S. Dupont, et al., 2004). As a research in state-of-the art, the research fields is expanded to speech communicationsbetween a patient and an operator in a Magnetic Resonance Imaging (MRI) room which has a noisy sound environment with a strong magnetic field (A. Moelker et al., 2005). Conventional microphonesuch as an accelerometercomposed of magnetic materials are not allowed in this environment, which requires a special microphone made of non-magnetic material.

For this environment the authors proposed a speech communication system that uses a BCS microphone with an optical fiber bragg grating (OFBG microphone) (M. Nakayama et al., 2011). It is composed of only non-magnetic materials, is suitable for the environment and should provide clear signals using our retrieval method. Previous research using an OFBG microphone demonstrated the effectiveness and performance of signal extraction in an MRI room. Its performance of speech recognition was evaluated using an acoustic model constructed with unspecified normal speech (M. Nakayama et al., 2011). It is concluded that an OFBG microphone can produce a clear signal with an improved performance compared to an acoustic model made by unspecified speeches. The original signal of an OFBG microphone enabled conversation however some stress was felt because its signal was low in sound quality. Therefore one of the research aims is to improve the quality with our retrieval method which used differential acceleration and noise reduction methods.

In this chapter, it will be shown in experiments and discussionsfor the body-conducted speeches with the method which is measured with an accelerometer and an OFBG microphone, as one of topics is a state-of-the-art in the research field of signal extraction under noisy environment.

Especially, it is mainly investigated in evaluations of the microphones, signal retrievals with the method and applying the method toa signal in sentence unit long for estimating and recovering of sound qualities.

2. SPEECH AND BODY-CONDUCTED SPEECH

2.1. Conventional Body-Conducted Speech Microphone

Speech as air-conducted sound is easily affected by surrounding noise. In contrast, body-conducted speech is solid-propagated sound and thus less affected by noise. A word is uttered by a 20-year-old male in a quiet room. Table 1 details the recording environments for microphone and acclerometer emploied in this research. Speech is measured 30 cm from the mouth using a microphone, and body-conducted speech is extracted from the upper lip using the accelerometer as conventional microphone which is shown in Figure 1. This microphone position is that commonly used for the speech input of a car navigation system. The upper lip, as a signal-extraction position, provides the best cepstral coefficients as feature parameters for speech recognition (S. Ishimitsu et al., 2004). Figures 2 and 3showuttered words "Asahi" in quiet room, taken from the JEIDA database, which contains 100 local place names(S. Itahashi, 1991). Speech is measured a cleary signal in frequency characteristics however body-conducted speech lacks high-frequency components above 2 kHz.So the performance is reduced when the signal is used for the recognition directory.

Table 1. Recording environments for microphone and accelerometer

Recorder	TEAC RD-200T
Microphone	Ono Sokki MI-1431
Microphone amplifier	Ono Sokki SR-2200
Microphone position	30cm (Between mouth and microphone)
Accelerometer	Ono Sokki NP-2110
Accelerometer amplifier	Ono Sokki PS-602
Accelerometer position	Upper lip

Figure 1. Accelerometer

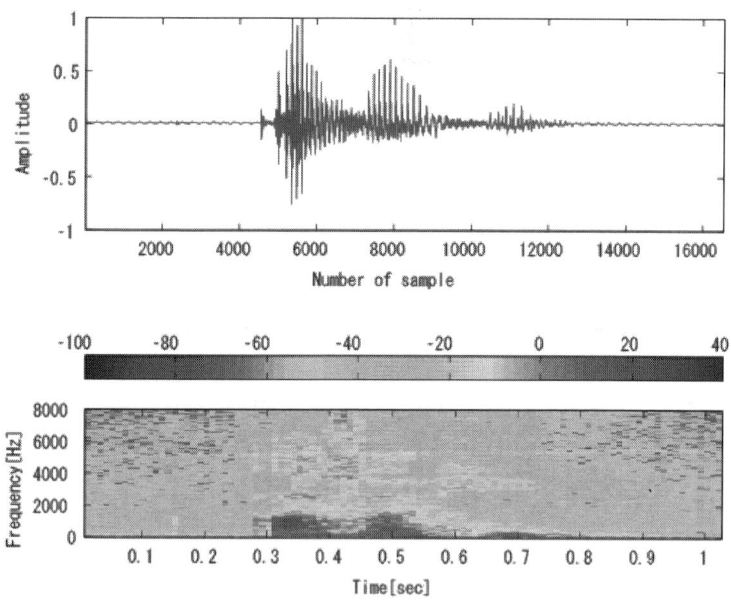

Figure 2. Speech from microphone in quiet

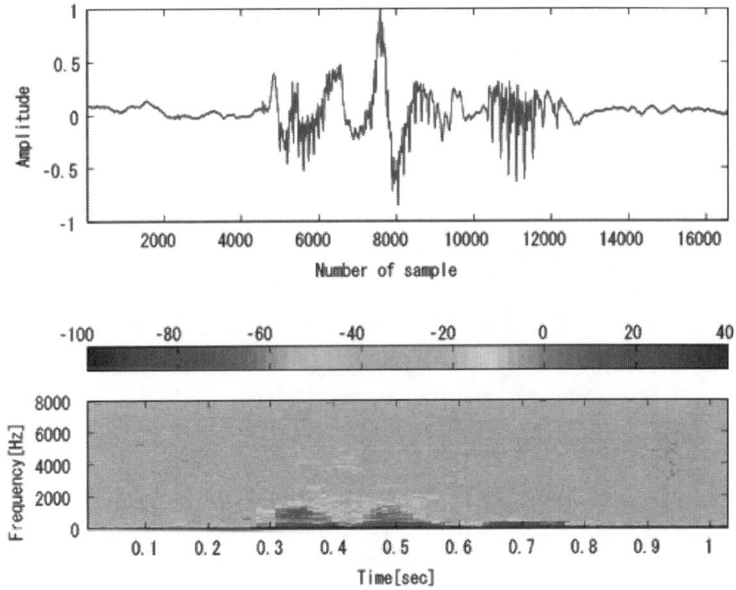

Figure 3. BCSfrom accelerometer in quiet

2.2. Optical Fiber Bragg Grating Microphone

To extend testing to scenarios such as that in which noise sound is generated with strong magnetic field, in communications between a patient and an operator in an MRI room, an OFBG microphone is employedto record body-conducted speech there because it can measure a clearer signal than an accelerometer and be used in an environment with a strong magnetic field. It is examined the effectiveness of the microphone in an MRI room in which a magnetic field is produced by an open-type magnetic resonance imaging system. Tables 2 and 3 detail the recording environments for OFBG microphone which is shown in Figure 4. Noise levels in the room did not measure at the recording point such as the mouth of the speaker because a sound-level meter did not permit into the room since it composed from magnetic materials. Therefore, the noise level is measured at the entrance of the room, and consequently may be higher than the noise level at the signal recording point; the noise level is given in Table 2. Owing to patient discomfort during the recordings, only 20 words and 5 sentences were recorded in the room where a scene is shown in

Figure 5. Figure 6 shows the body-conducted speech recorded from the OFBG microphone in the room when activated a MRI. Compared the signal with conventional BCS, it is clearer than that for body-conducted speech measured by accelerometer because characteristics of frequencies above 2 kHz can be found.

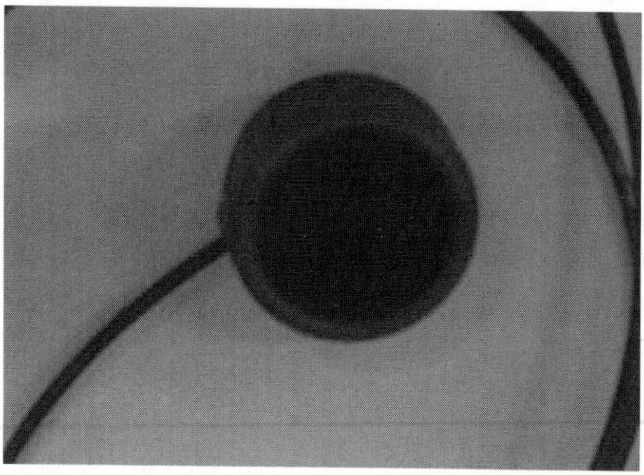

Figure 4. OFBG microphone

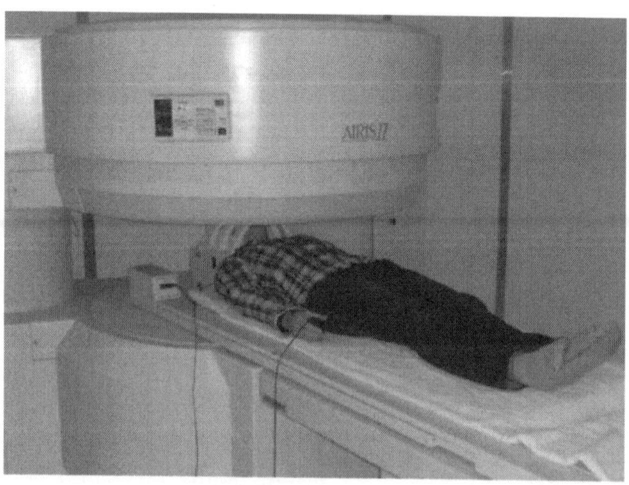

Figure 5. Signal recording in an MRI room

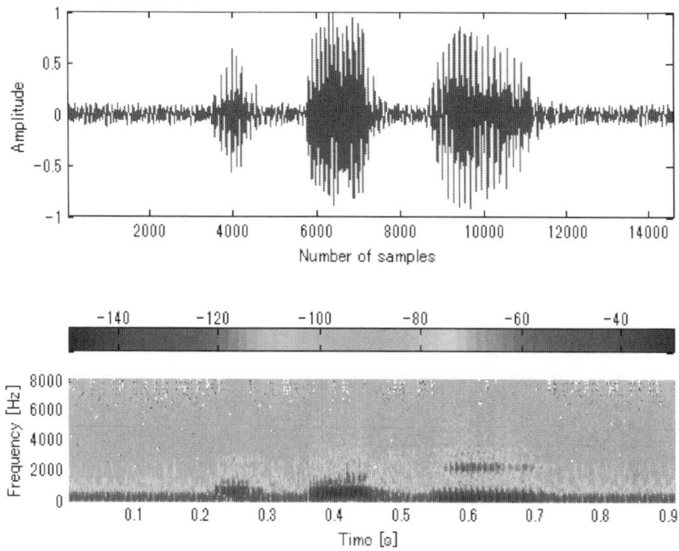

Figure 6. BCS from OFBG microphone

Table 2. Recording environment 1 for OFBG microphone

MRI model	HITACH AIRIS II
Environment	MRI (OFF): 61.6 dB SPL MRI (ON): 81.1 dB SPL
Speakers	two males (22 and 23 years old) two females (23 and 24 years old)
Vocabulary	twenty words × two sets: JEIDA 100 local place names five sentences × three sets: ATR database sentences

Table 3. Recording environment 2 for OFBG microphone

Device name	Type name
Pickup	Optoacoustics Optimic4130
Optical-electronic conversion device	Optoacoustics EOU200
Recorder	TEAC LX-10

3. SPEECH RECOGNITION WITH OFBG MICROPHONE

The quality of the signal recorded with the OFBG microphone, is higher than the quality of BCS recorded with accelerometer. Generally, the quality of speech sound is evaluated by the mean opinion score from 1 to 5 however this requires much evaluation data to achieve adequate significance levels. For the reason, it is evaluated the sound quality through speech recognition using acoustic models estimated with the speech of unspecified speakers as results of recognition performances. In speech recognition, the best candidate is chosen and decided by likelihoods derived from acoustic models and feature parameters such as cepstral parameters, which are calculated from the recorded speech (D. Li, and D. O'Shaughnessy, 2003) (L. Rabiner, 1993). As a result, the recognition performances and likelihoods are statistical results since human errors and other factors are not considered.

3.1. Experimental Conditions

Table 4 shows the experimental conditions for isolated word recognition in speech recognition. The experiment employs the Julius, speech recognition decoder, which is a large-vocabulary continuous-speech recognition system for Japanese language (T. Kawahara et al., 1999) (A. Lee et al., 2001). The decoder requires a dictionary, acoustic models and language models. The dictionary describes connections of sub-words in each word, such as phonemes and syllables, which are the acoustic models. Language models give the probability for a present word given a former wordin corpora. The purpose of the experiment is only the evaluation of the clarity or the similarity of signals and acoustic models. Since language models are not required in this experiment, Julian version 3.4.2 is used for isolated-word recognition especially. Thus, the experimentsare used the same acoustic models estimated by HTK with JNAS to evaluate closeness of signals when highest recognition performance is achieved (S. Young et al., 2000) (K. Itou et al, 1999).

3.2. Experimental Results

Table 5 shows recognition results of isolated word recognition in each data set, and Table 6 gives averages of recognition results in each speaker. The recognition results for the OFGB microphone are found to be superior to the recognition results for the conventional BCS microphone. The differences in isolated-word recognition rates are about 15% to 35% respectively. These results show the effectiveness of the OFBG microphone when is measured clearly signals with it.

Table 4. Experimental conditions for isolated word recognition

Speaker	two males (22 and 23 years old) two female (23 and 24 years old)
Number of datasets	20 words × three sets/person
Vocabulary	JEIDA 100 local place names
Recognition system	Julian 3.4.2
Acoustic model	gender-dependent triphone model
Model conditions	16 mixture Gaussian, clustered 3000 states
Feature vectors	MFCC(12)+ΔMFCC(12)+ΔPow(1)=25 dim.
Training condition	more than 20,000 samples JANS with HTK 2.0

Table 5. Recognition results of isolated word recognition in each data set

Speaker	MRI off			MRI on		
	set 1	set 2	set 3	set 1	set 2	set 3
Male 1	85%	80%	90%	30%	40%	50%
Male 2	90%	75%	85%	50%	60%	60%
Female 1	35%	35%	35%	20%	20%	20%
Female 2	80%	70%	70%	75%	70%	75%

Table 6. Averagesof recognition results

Speaker	MRI off	MRI on
Male 1	85.0%	40.0%
Male 2	83.3%	56.7%
Female 1	35.0%	20.0%
Female 2	73.3%	73.3%

4. IMPROVEMENT ON SOUND QUALITY OF BODY-CONDUCTED SPEECH IN WORD UNIT

The OFBG microphone can measure a high quality signal compared to a BCS of an accelerometer. To realize conversations without stress, signals with improved in sound qualitiesare required. Consequently, one of aims in the research is to invent and examine a method for improving sound quality.Many researchers and researches which are already introduced in the chapter of introduction, are unaware that a BCS does not have frequency components 2 kHz and higher. Mindful of this condition, conventional retrieval methods for BCS that need the speech and its parameters are proposed and investigated, however speech is not measured easily in noisy environments. Therefore a signal retrieval method for a BCS only performs well with itself. In realizing this progressive idea, the method is invented a signal retrieval method without speech and the other parameters becauseeffective frequency components in signals over 2 kHz are found however there contains very low gains.

4.1. Differential Acceleration

Formula (1) shows an equation for estimating using the differential acceleration from the original BCS.

$$x_{differential}(i) = x(i+1) - x(i) \tag{1}$$

xdifferential(i) is the differential acceleration signal that is calculated from each frame of a BCS. Because of low gains in its amplitude, it requires adjusting to a suitable level for hearing or processing. Figure 7 shows a differential acceleration estimated from Figure 6 using Formula (1), with the adjusted gain. It seems that the differential acceleration signal is composed of speech mixed with stationary noise, so we expected to be able to remove it completely with the noise reduction method because the signal has a high SNR compared to the original signal. Consequently, it isproposed the signal estimation method using differential acceleration and a conventional noise reduction method (M. Nakayama et al., 2011).

4.2. Noise Reduction Method

As a first approach to noise reduction, it is examined the effectiveness of a spectral subtraction method for the reduction of stationary noise. However, improvements in performancesfor the frequency components is inadequated with this approach. The noise spectrum is simply subtracted by a spectral subtraction method, so a Wiener-filtering method is expected to estimate the spectrum envelope of speech using linear prediction coefficients. Therefore, it is tried to extract a clear signal using the Wiener-filtering method, which could estimate and obtain the effective frequency components from noisy speech. Formula (2) shows the equation used for the Wiener-filtering method.

$$H_{Estimate}(\omega) = \frac{H_{Speech}(\omega)}{H_{Speech}(\omega) + H_{Noise}(\omega)} \qquad (2)$$

An estimated spectrum *HEstimate(ω)* can be converted to a retrieval signal from the differential acceleration signal. It can be calculated from the speech spectrum *HSpeech(ω)* and noise spectrum*HNoise(ω)*. In particular, *HSpeech(ω)* is calculated with autocorrelation functions and linear prediction coefficients using a Levinson-Durbin algorithm (J. Durbin, 1960), and *HNoise(ω)* is then estimated using autocorrelation functions.

4.3. Evaluations

Signal retrieval for a signal measured by an OFBG microphone is performed using the same parameters in the method because a propagation path of body-conducted speech in a human body is not affected by either quiet or noisy environments. Figure 8 shows a retrieval signal from Figure 7 using a Wiener-filtering method where the linear prediction coefficients and autocorrelation functions are 1 and the frame width is 764 samples. These procedures were repeated five times on a signal to remove a stationary noise. From a retrieval signal, high frequency components from 2 kHz and above were recovered with these settings. This proposed method could also be applied to obtain a clear signal from body-conducted speech measured with OFBG microphone in noisy sound and high magnetic field environment.

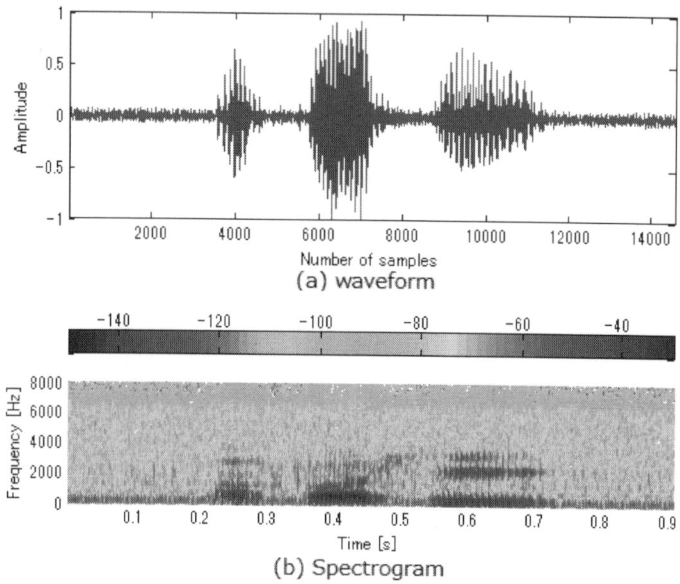

Figure 7. Differential acceleration from OFBG microphone

5. IMPROVEMENT ON SOUND QUALITY OF BODY-CONDUCTED SPEECH IN SENTENCE UNIT

The effectiveness of signal retrieval for body-conducted speech in word unit measured by an accelerometer and an OFBG microphone has been demonstrated at former sections. Howeverthe effectiveness of body-conducted speech in word unit is proven,signals in sentence unit need to be examined for practical use such as conversations in the noisy environment. Though the investigation for the sentence unit is an important evaluation, so it could revolutionize speech communications in the environment. As a first step in signal retrieval for sentence unit, themethod adoptsthe method to signals in word unit because the transfer function between the microphone and sound source seems to change little whether word or sentence unit, and is examineda body-conducted speech insentence unit directly measured by an accelerometer and an OFBG microphone.

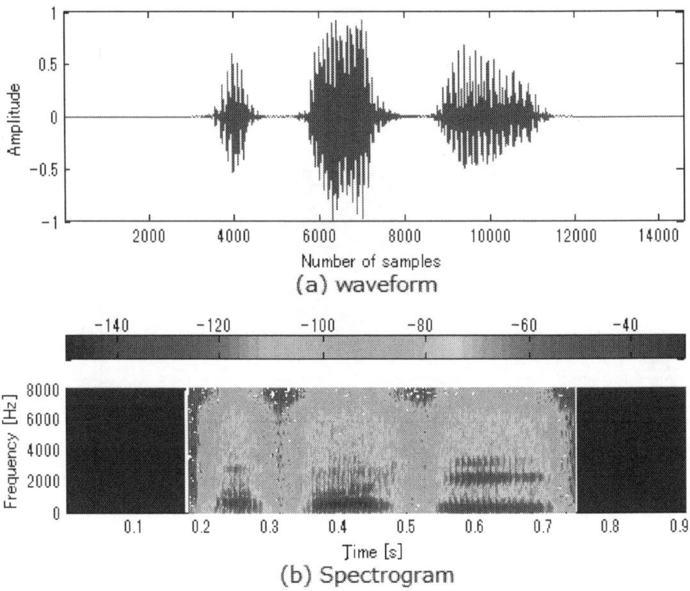

Figure 8. Retrieval signal from OFBG microphone

5.1. Body-Conducted Speech From An Accelerometer

In experiments on signal retrieval using an accelerometer, speech and body-conducted speech were measured in a quiet room of our laboratoryand engine room of the training ship at the Oshima National College of Maritime Technology, where there is noisy environments with working a main engine and two generator, are shown Figures 9 (a) and (b). The recording environment is also usedTable 1, however the speaker who uttered a word differs from a speaker in a former section. Noise within the engine room, under the two conditions of anchorage and cruising, were 93 and 98 dB SPL, respectively, and the SNR measurements from microphone. There was –20 and –25 dB SNR, respectively. In this research, the signal is experimented under cruisingcondition to estimate retrieval signals.

A 22-year-old male uttered A01 sentencefrom the ATR503 sentence database, andthe sentence is a commonly used sentence in speech recognition and application (M. Abe et al., 1991). And the sentence is composed of the followings in sub-word of mora.

/a/ /ra/ /yu/ /ru/ /ge/ /N/ /ji/ /tsu/ /wo/ /su/ /be/ /te/ /ji/ /bu/ /N/ /no/ /ho/ /u/ /he/ /ne/ /ji/ /ma/ /ge/ /ta/ /no/ /da/

(a)Main engine of Oshima-maru

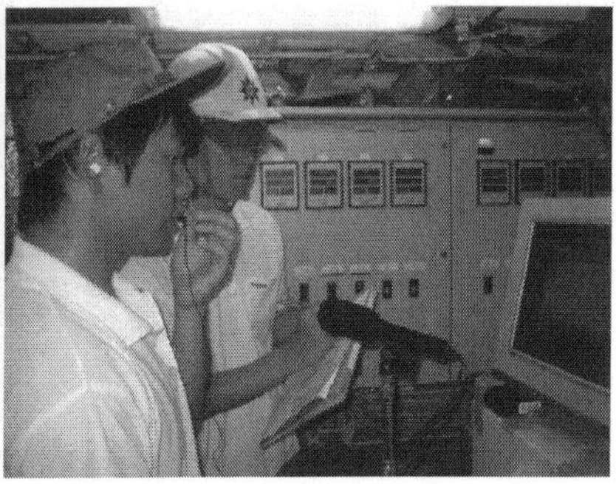

(b)Signal recording in the engine room

Figure 9. The engine room in Oshima-maru

Figures10 and 11 show a speech and a body-conducted speech insentence unit measured by a conventional microphone and accelerometerin a quiet room when a 22 years-old male uttered the sentence. Although the accelerometeris held with fingers, soundsare measured clearly because it was firmly held to the upper lip with a suitablepressure. Figure 12 shows a differential acceleration fromFigure 11, becomes clearly signal with little noise because the BCS is high SNR.

Figures 13 and 14 show a speech and a body-conducted speech in sentence unit in the noisy environment. Speech is completely swamped by the intense noise from the engine and generators. On the other hand, body-conducted speech in Figure 14 is affected a little by the noise but can be measured. Because SNR in Figure 14 has low gain, differentialacceleration in Figure 15 is considered that the performance of signal retrieval is reduced. Figure 16 shows the signal retrieval from the differential accelerationworks well when the treated four times since the performance is sufficient to recover the frequency characteristics. As a result, it is concluded that body-conducted speech is as clear as possible without noise disturbance.

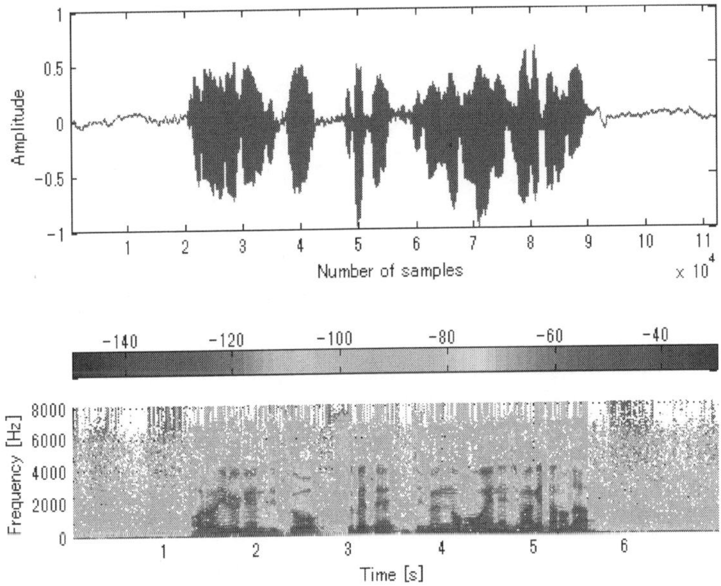

Figure 10. Speech of sentence in quiet

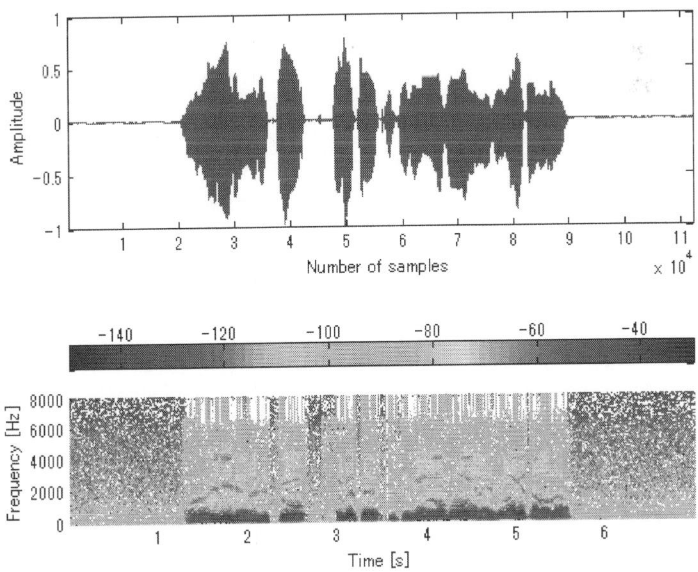

Figure 11. BCS of sentence in quiet

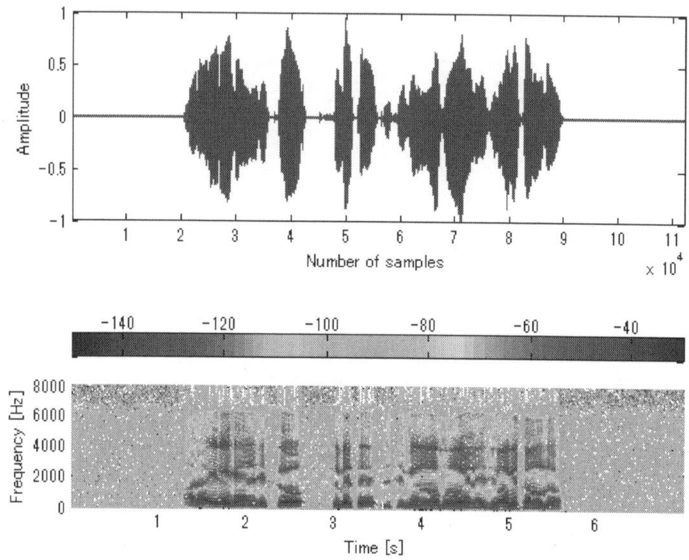

Figure 12. Differential accelerationof sentence in quiet

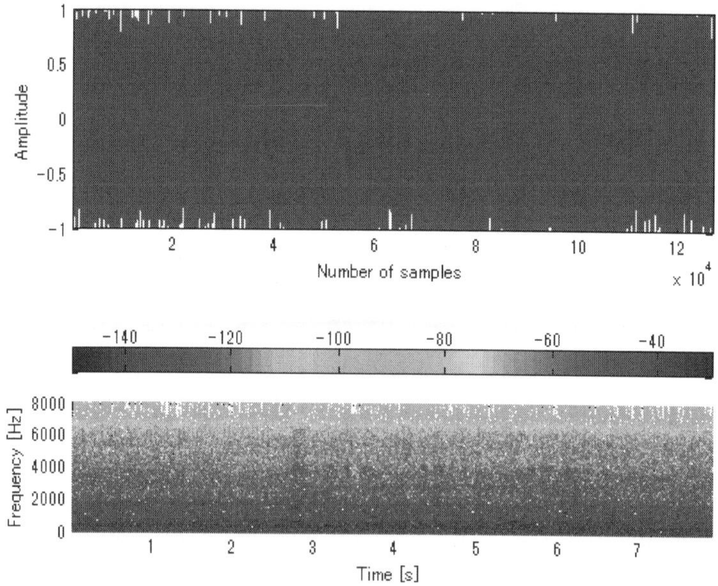

Figure 13. Speech of sentence innoise environment

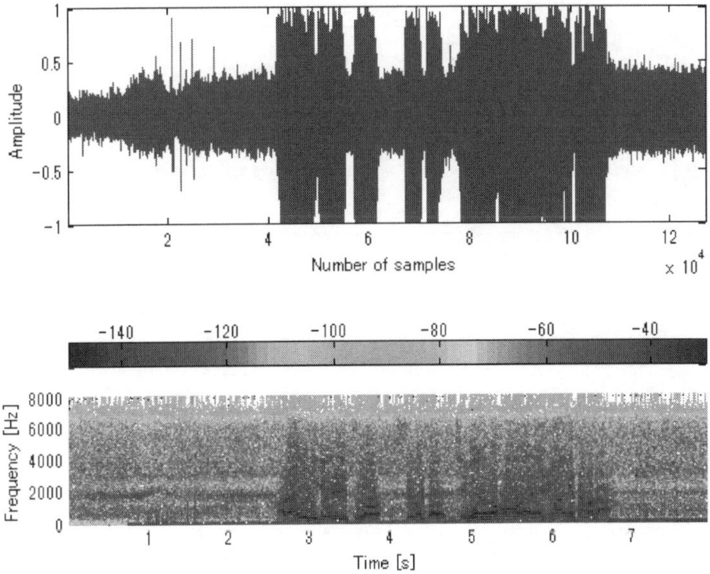

Figure 14. BCS of sentence in noise environment

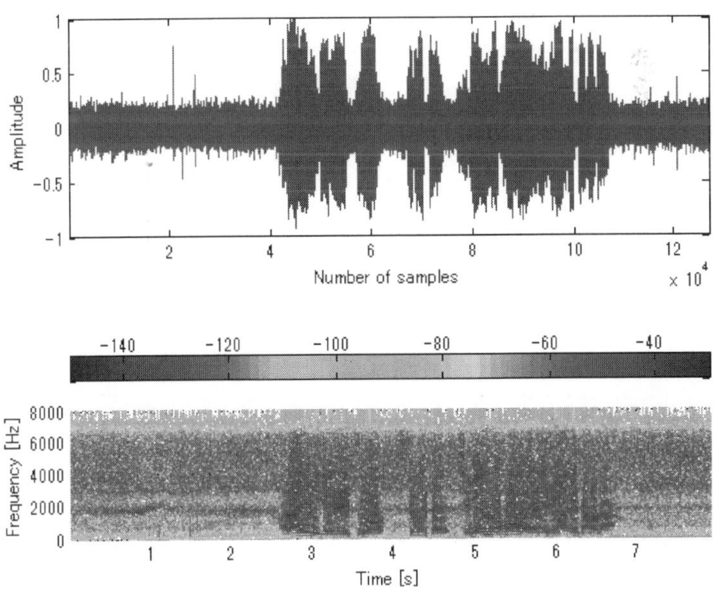

Figure 15. Differential acceleration of sentence in noise environment

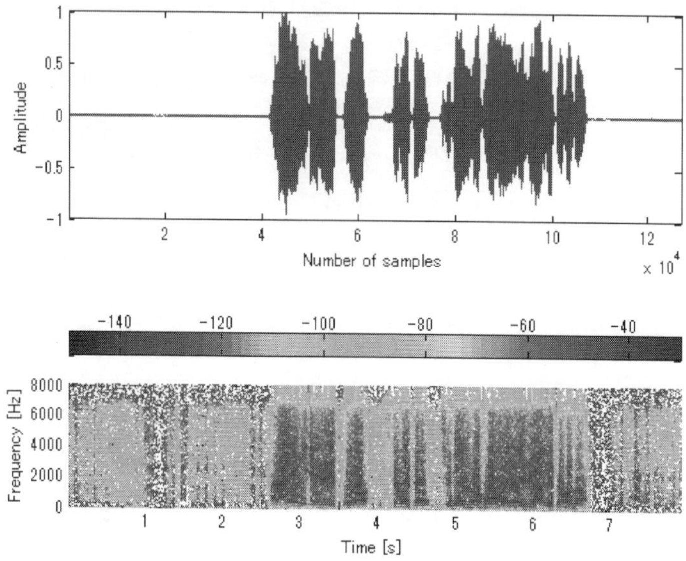

Figure 16. Retrieval BCS of sentence in noise environment

5.2. Body-Conducted Speech From Ofbg Microphone

The quality of the signal measured by the OFBG microphone in the noisy environment of an MRI room was investigated here. A speaker uttered the sentence A01 during the operation of MRI devices, such that there was an 81 dB SPL-noise environment. Although a sound level meter was not permitted in the room, so itis measured in front of the gate door in the room.Figure 17 shows the signal of the uttered sentence recorded by the OFBG microphone in the MRI room when MRI equipment was in operation. Since the signal is clear, it is expected that the frequency characteristics of the signal can be recovered employing the signal retrieval method. Figures 18 and 19 show the differential acceleration and retrieved signal from the OFBG microphone in the MRI room when the MRI equipment was in operation and the method treated three times. These figures confirm to improve in the sound quality of BCS in sentence, and it also concluded that the SNR in BCS is best when it has high level.

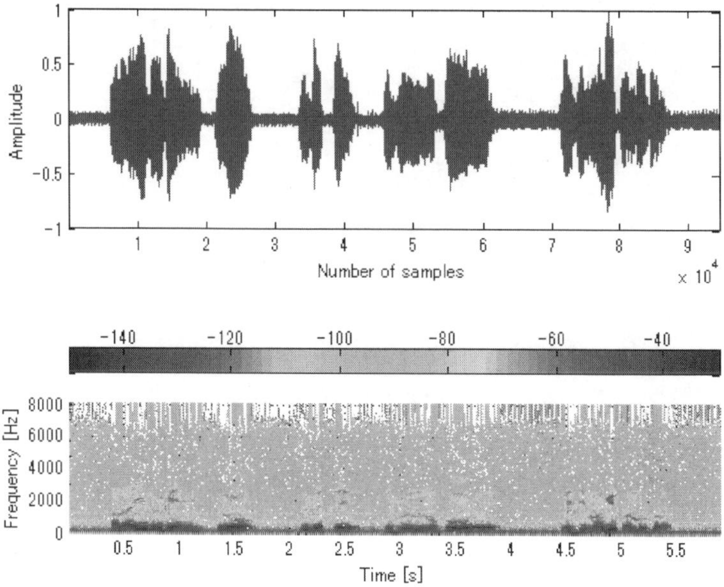

Figure 17. BCS of sentence in MRI room

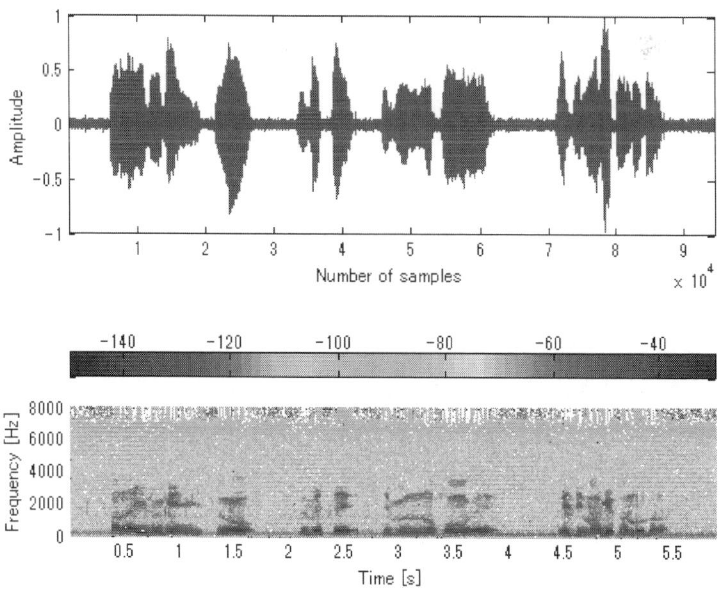

Figure 18. Differential acceleration of sentence in MRI room

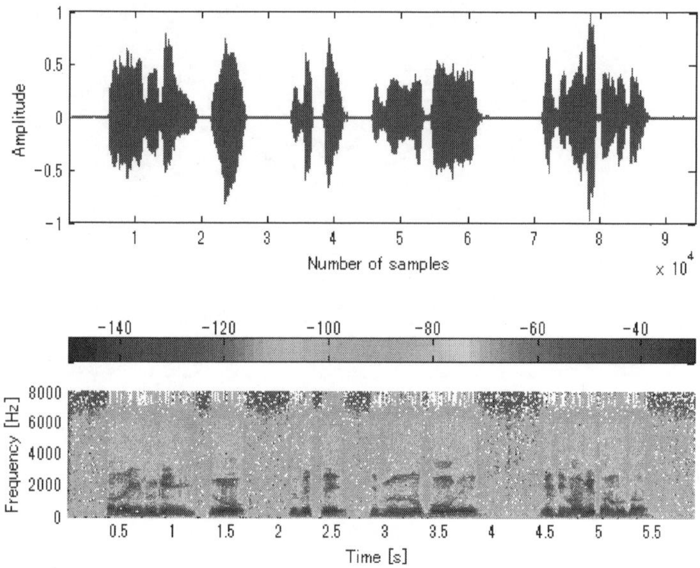

Figure 19. Retrieval signal of sentence in MRI room

6. CONCLUSIONS AND FUTURE WORKS

This section presentsimprovements on sound quality of body-conducted speeches measured with an accelerometer and an OFBG microphone. Especially, an MRI room has heavy noisy sound and high magnetic field environment. The environment does not allow bringingaccelerometer such as a conventional body-conducted speech microphone which is made from magnetic materials. For conversations and communications between a patient and an operator in the room, an OFBG microphone is proposed, which can measure clear signals compared to accelerometer.

And then, the performances of signal retrieval method in sentence with the microphones that are an accelerometer and an OFBG microphone were evaluated, and the effectiveness is confirmed with time–frequency analysis and speech recognition. From this background, it is investigated estimating clear body-conducted speech in sentence unit from an OFBG microphone with our signal retrieval method that used combined differential acceleration and noise reduction. Applying the method to the signal measured recovered which in sound quality that was evaluated using time-frequency analysis. Thus, its retrieval method can also be applied to a signal measured by an OFBG microphone with the same settings because its conduction path is not affected by the noise in the air. The signals were measured in quiet and noisy rooms, specifically an engine room and MRI room. The signals were clearly obtained employing the signal retrieval method and the same settings used for the word unit as a first step. To obtain a clearer signal with the signal retrieval method, the pressure at which the microphone is held is important, and the sounds have high SNR in original BCS.

As future works, it needs to extend the signal retrieval method for practical use and improvement of algorithm for advance.

ACKNOWLEDGEMENT

The authors thank Mr. K. Oda, Mr. H. Nagoshi and his colleagues in Ishimitsu laboratory of Hiroshima City University, members of the Living Informatics Research Group, Health Research Institute, National Institute of Advanced Industrial Science and Technology (AIST) for their support in the signal recording, and crew members of the training ship, Oshima-maru, Oshima National College of Maritime Technology.

REFERENCES

1. A. Lee, T. Kawahara, K. Shikano, 2001Julius- an open source real-time largevocabulary recognition engine, in Proceedings of European Conference on SpeechCommunication and Technology (EUROSPEECH), 16911694

2. A. Moelker, R. A. J. J. Maas, M. W. Vogel, M. Ouhlous, P. M. T. Pattynama, 2005Importance of bone-conducted sound transmission on patient hearing in the MR scanner, Journal of Magnetic Resonance Imaging, 221163169

3. D. Li, D. , O'Shaughnessy(2003Speech Processing: ADynamic and Optimization-Oriented Approach, MarcelDekker Inc.

4. H. Hirsch, D. Pearce, 2000The AURORA experimental framework for the performance evaluation of speech recognition systems under noisy conditions, in proceedings of ISCA ITRW ASR2000, 181188

5. J. Durbin, 1960The Fitting of Time-Series Models, Review of the International Statistical Institute, 283233244

6. K. Itou, M. Yamamoto, K. Takeda, T. Takezawa, T. Matsuoka, T. Kobayashi, K. Shikano, S. Itahashi, 1999JNAS : Japanese speech corpus for large vocabulary continuous speech recognition research, Journal of the Acoustical Society of Japan (E), 20(3), 199206

7. L.Rabiner(1993Fundamentals of Speech Recognition, PrenticeHall

8. M. Abe, Y. Sagisaka, T. Umeda, H. Kuwabara, 1990Manual of Japanese Speech Database, ATR

9. M. Nakayama, S. Ishimitsu, S. Nakagawa, 2011A study of making clear body-conducted speech using differential acceleration, IEEJ Transactions on Electrical and Electronic Engineering, 62144150

10. M. Nakayama, S. Ishimitsu, H. Nagoshi, S. Nakagawa, K. Fukui, 2011Body-conducted speech microphone using an Optical Fiber Bragg Grating for high magnetic field and noisy environments, in proceedings of Forum Acusticum 2011

11. N. Kitaoka, T. Yamada, S. Tsuge, C. Miyajima, T. Nishiura, M. Nakayama, Y. Denda, M. Fujimoto, K. Yamamoto, T. Takiguchi, S. Kuroiwa, K. Takeda, S. , Nakamura(2006CENSREC-1-C: development of evaluation framework for voice activity detection under noisyenvironment, IPSJ SIG Technical Report, 2006-SLP-63,16

12. S. Dupont, C. Ris, D. Bachelart, 2004Combined use of closetalk and throat microphones for improved speech recognition under non-stationary background noise, in proceedings of COST278 and ISCA Tutorial and Research Workshop (ITRW) on Robustness Issues in Conversational Interaction, paper31

13. S. Ishimitsu, 2008Construction of a Noise-Robust Body-Conducted Speech Recognition System, in Chapter of Speech Recognition, IN-TECH

14. S. Ishimitsu, H. Kitakaze, Y. Tsuchibushi, H. Yanagawa, M. Fukushima, 2004A noise-robust speech recognition system making use of body-conducted signals, Acoustical Science and Technology, 252166169

15. S. Ishimitsu, M. Nakayama, Y. Murakami, 2004Study of Body-Conducted SpeechRecognition for Support of Maritime Engine Operation, in Journal of the JIME,3943540in Japanese)

16. S. Itahashi, 1991A noise database and Japanese common speech data corpus, Journal of ASJ, 4712951953

17. S. Young, J. Jansen, J. Odell, P. Woodland, 2000The HTK Book for 2CambridgeUniversity

18. T. Kawahara, A. Lee, T. Kobayashi, K. Takeda, N. Minematsu, K. Itou, A. Ito, M. Yamamoto, A. Yamada, T. Utsuro, K. Shikano, 1999Japanese dictation toolkit- 1997version, Journal of ASJ, 203233239

19. T. T. Vu, M. Unoki, M. Akagi, 2006A Study on Restoration of Boneconducted Speech With LPC Based Model, IEICE TechnicalReport, SP2005-174, 6778

20. T. Tamiya, T. Shimamura, 2006Improvement of Body-Conducted Speech Quality by Adaptive Filters, IEICE Technical Report, SP2006-191, 4146

21. Z. Liu, Z. Zhang, A. Acero, J. Droppo, X. Huang, 2004Direct Filtering for Air- and Bone-Conductive Microphones, in proceedings of IEEE International Workshop on Multimedia Signal Processing (MMSP'04), 363366

Chapter 6

Stand-Alone Intelligent Voice Recognition System

Mohammed R. Saady[1], Hatem El-Borey[2], El-Sayed A. El-Dahshan[2,3], Ashraf Shamseldin Yahia[2]

[1]Basic Science Department, Faculty of Computers and Information, Fayoum University, El Fayoum, Egypt

[2]Physics Department, Faculty of Science, Ain Shames University, Cairo, Egypt

[3]Egyptian E-Learning University, Elmesaha El-Doki, Giza, Egypt

ABSTRACT

In this paper, an expert system for security based on biometric human features that can be obtained without any contact with the registering sensor is presented. These features are extracted from human's voice, so the system is called Voice Recognition System (VRS). The proposed system consists of a combination of three stages: signal pre-processing, features extraction by using Wavelet Packet Transform (WPT) and features matching by using Artificial Neural Networks (ANNs). The features vectors are formed after two steps: firstly, decomposing the speech signal at level 7 with Daubechies 20-tap (db20), secondly, the energy corresponding to each WPT node is calculated which collected to form a features vector. One hundred twenty eight features vector for each speaker was fed to the Feed Forward Back-propagation Neural Network (FFBPNN). The data used in this paper are drawn from the English Language Speech Database for Speaker Recognition (ELSDSR) database which composes of audio files for training and other files for testing. The performance of the proposed system is evaluated by using the test files. Our results showed that the rate of correct recognition of the proposed system is about 100% for training files and 95.7% for one testing file for each speaker from the ELSDSR database. The proposed method showed efficiency results were better than the well-known Mel Frequency Cepstral Coefficient (MFCC) and the Zak transform.

KEYWORDS

Voice Recognition, Wavelet Packet Transform, Feature Extraction, Artificial Neural Network

1. INTRODUCTION

Everywhere around us are signals that can be analyzed. Signals are time-varying quantities which carry a lot of information. They may be audio signals, images or video signals, sonar signals or ultrasound, biological signals such as the electrical pulses from the heart, communications signals, or many other types. Audio signals carry an enormous amount about you.

In the last several years, several numbers of security systems based on fingerprints, voice, iris, face images, the pattern of the hands, and the pattern of motion of the arms have been presented. The security system based on iris or fingerprint is not convenient in practice since it imposes stringent requirements on interaction with the user; the users of modern fingerprint become anxious about the hygiene of the process, and the users of modern iris face stringent requirements on movements and visibility of the eyes [1] . So there is a strong motivation to work in security system based on voice.

The security systems based on voice have been increased rapidly because of its non-contact characteristic [2] , and no two individuals sound identical since their vocal tract shapes, larynx sizes, and other parts of their voice production organs are different [3] . So the Voice Recognition System (VRS) became one of the most useful and popular biometric recognition systems in the world especially in the areas in which security was a major concern. It can be used for authentication, surveillance, forensic speaker recognition and a number of related activities.

Voice Recognition (VR) which also called Speaker Recognition (SR) is the process of automatically recognizing who is speaking depending on the basis of individual information included in his/her voice signal. SR can be classified into Speaker Identification (SI) and Speaker Verification (SV). Speaker Identification is the process of determining which registered speaker provides a given utterance (one-to-many matching), on the other hand, SV is the process of accepting or rejecting the identity claim of a speaker (one-to-one matching), and this decision depends on the degree of similarity between the claimed speaker and the enrolled speaker.

In VRS, two basic operations are carried out: feature extraction and classification. In feature extraction, we capture the vital characteristic which enables us to distinguish a speaker from the other. Some of techniques were used for extracting features were Fourier Transforms (FTs), and Short-Time Fourier Transforms (STFTs). But these techniques aren't suitable for representing non stationary signals such as voice signals. By using (FTs), the input signal was transformed from the time domain into the frequency domain so we could know the frequencies that contained in the signal but not when, and STFTs tried to handle the problem by using a local window to display the frequency and time relation but its accuracy depended on the used window. Wavelet Transforms (WTs) have handled some of these problems. Classification is executed at two steps: enrollment and matching which often referred to them as training and testing respectively. In enrollment step, the speaker is introduced to the system by using the extracted features from the training data, and the testing step is activated when a vector of data from unknown speaker is entered to the system. It matches the vector data to a model corresponding to a known speaker. Classifications techniques have been used in VRS include Hidden Markov Models (HMMs) and Artificial Neural Networks (ANNs).

Several studies based on WTs as well as ANNs have been presented to design a SI system [2] [4] . A text-in- dependent speaker identification system based on the Zak transform [5] has been presented where the speech database was drawn from the ELSDSR database [6] . In this study, we have tried to enhance the efficiency of the SI System which has already existed [5] . This enhancement can be summarized as: Wavelet Packet Transform (WPT) has been applied at the stage of feature extraction but these data were not suitable for classification as the performance of ANNs is depended on the size of the training data. The critical choice of subset of features from a larger set is very important to improve the performance of speaker recognition [7]. In our study, the energy corresponding to WP nodes have been captured as features vectors. These features vectors have been fed to Feed Forward Back-propagation Neural Network (FFBPNN) to train it. Finally, we have tested the trained FFBPNN with a single test file for each speaker.

The rest of the paper is organized as follows: Section 2 gives in brief an overview on the structure of a general VRS. Section 3 introduces the proposed VRS. In Section 4 the results and discussions were presented. Finally, Section 5 presented a conclusion and the future work.

2. GENERIC VOICE RECOGNITION SYSTEM

Each VRS includes two phases: training and testing phases as graphed in Figure 1. In the training phase the speaker's voice print is set up from the features vector. These voice prints are stored in a reference database. While in the testing phase, features are extracted from unknown speaker's voice, then a pattern matching algorithm computes the similarity score between the unknown speaker's features vector and that stored in the reference database. Both the training and the testing phases include feature extraction so it is often called the front- end of the system.

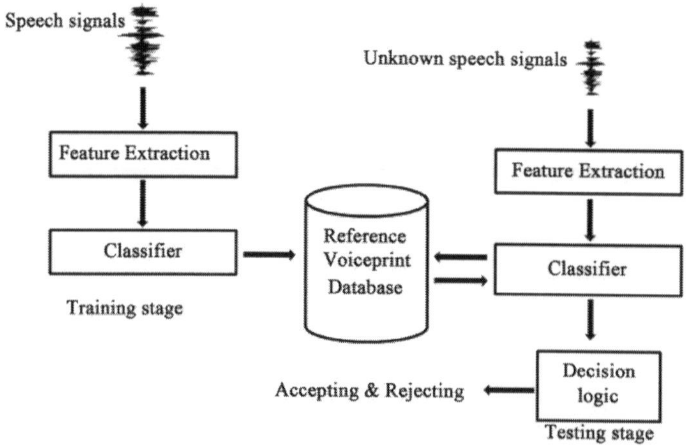

Figure 1. Typical voice recognition system.

2.1. Feature Extraction

Here we extract amount of features from the voice signal. The extracted features should be capable of separating the speakers from each other in its space. In traditional techniques speech features are extracted by FTs and (STFTs) [2] . But these techniques are not suitable for representing voice because they accept stationary signals within a given time frame therefore lack the ability of these techniques to analyze the non-stationary signals or signals in transient state [8] . One of the features obtained via FTs and used in the recognition task was the Mel Frequency Cepstral Coefficient (MFCC). But the recent researches showed that the identification rate with MFCCs can be as high as 99.5% for the noise free TIMIT database [9] . However, the identification accuracy reduced to 60% for the same data set that was transmitted over telephone channels. In 1984, The wavelet theory was proposed, Goupillaud et al. introduced a new transformation for the frequency analysis of the discretized signals. The transform is known as Wavelet Transform (WT) [10] [11] . It gives us the ablllty for decomposing any signal (analog or digital) of any field (medical, biometric, etc.) into different resolution levels and displays two important variables (frequency & time)synchronously. Wavelets can be applied to many types of problems such as signal de-noising, compression and feature extraction.

The analysis by the Wavelet Transforms is performed by using a single wave is called a mother wavelet (often called window) Ψ(t) which can be considered as a band pass filter, has a limited duration of zero average, has irregular form, and often non-symmetrical unlike sinusoidal waves which extends from minus to plus infinity, and has a symmetrical form. The wavelet transform is defined as the inner product of a signal x(t) with the mother wavelet Ψ(t).

$$\Psi_{a,b}(t)=\psi\left(\frac{t-b}{a}\right)$$

$$W_{\Psi}x(a,b)=\frac{1}{\sqrt{a}}\int_{-\infty}^{\infty}x(t)\psi^{*}\left(\frac{t-b}{a}\right)$$

where a, b are the scale and translation parameters.

Depending on the scale factor a, the Wavelet Transforms have a large freedom degree to vary window size. This varying window size is wide for slow frequencies because the low frequency component completes a cycle in a large time interval, and it is narrow for the other high frequencies as the high frequency component completes a cycle in a much shorter interval [12]. Thus, an optimal time-frequency resolution is obtained in all frequency ranges.

2.2. Classification

After the features are extracted and saved for each speaker, these values are considered the threshold to enter the second operation of recognition, classification, which means identification of the identity of unknown speaker by comparing the extracted features from his/her voice with the saved features (Reference Voiceprint Database), the reference which produces a maximum score of similarity is selected to yield a prediction for the speaker's identity. There are many classification tools such as Hidden Makrov Models (HMMs), Gaussian mixture model (GMM), and ANNs which has an increasing interests in classification and used in this study. The advantages of ANNs for solving speech/speaker recognition problems are their error tolerance and non-linear property [13] .

The ANNs is a mathematical model composed of a large number of simple processing units called neurons. The strategy of its work is inspired from the biological nervous system. Neurons are very lower than silicon logic gates, so the brain overcomes the relatively slow rate of operation by having a staggering number of neurons with massive interconnection between them. Neurons are connected to each other via synapses, and the strength of it is determined by the weight value it possess, each neuron as graphed in Figure 2 has an adder for summing the weighted input signal, and the output of the neuron is limited by the neuron's activation function which the total input of it is controlled by an externally applied bias.

$$U_k = \sum_{j=1}^{m} w_{kj} x_j$$

where w_{kj} is the synaptic weight directed from neuron j to neuron k, j is the number of inputs.

$$y_k = \Phi(U_k + b_k)$$

where Φ is the chosen activation function of the neuron, b_k is the bias.

These neurons are arranged in layers one after the other, and the network architecture lies in one of three different classes; single-layer networks where we have an input layer of source nodes that projects onto an output layer of neurons, multilayer networks where there is one or more hidden layer, and the third class is the recurrent networks where there is at least one feedback loop.

Applications of ANNs included; pattern association, pattern recognition, function approximation, fitting, and beam forming. Our study belongs to pattern recognition task.

3. THE PROPOSED VOICE RECOGNITION SYSTEM

The proposed system has been presented here is characterized by existing an operation before the feature extraction operation which included in any general VRS. This operation is called pre-processing, the purpose of it is to prevent the error estimation caused by speakers' volume changes, as see in Figure 3.

3.1. Database Used in This Study

The performance of the VRS which based on a standard speech database is comparable. There are 36 databases among free license such as the database which used here and proprietary bases that have been used in speaker recognition studies. The speech signals have been used in this study are drawn from the standard ELSDSR database. It is non-native speakers' database. It has been designed to provide speech data for the development and evaluation of automatic VRS. It has been designed by the Ph. D students and Master students from department of Informatics and Mathematical Modeling (IMM) at Technical University of Denmark. The speech language is English, and spoken by 21 Dane, one Islander and one Canadian. It contains voice messages from 23 speakers (13M/10F), and the age covered from 24 to 63. No a priori control of the speaker distribution by nationality and age has been done, except for the gender. Since the speakers were selected only in IMM, the speaker group exhibits relatively small variation in profession and educational background. The subjects of this database were from different countries and different places of one country, the dialect of reading English language in this database can probably be used as accent recognition. Part of the text, which is suggested as training subdivision, was made with the attempt to capture all the possible pronunciation of English language, which includes the vowels, consonants and diphthongs. With the suggested training and test subdivision, seven paragraphs of text are constructed and collected for training, which includes 11 sentences; and 46 sentences were collected for test text. In other word, for the training set, 161 (7 *23) utterances were recorded; and for test set, 46 (2 *23) utterances were provided. On average, the duration for reading the training data is: 78.6 s for male; 88.3 s for female; and 83 s for all. The duration for reading test data, on average, is: 16.1 s (male); 19.6 s (female); and 17.6 s (for all) [6] .

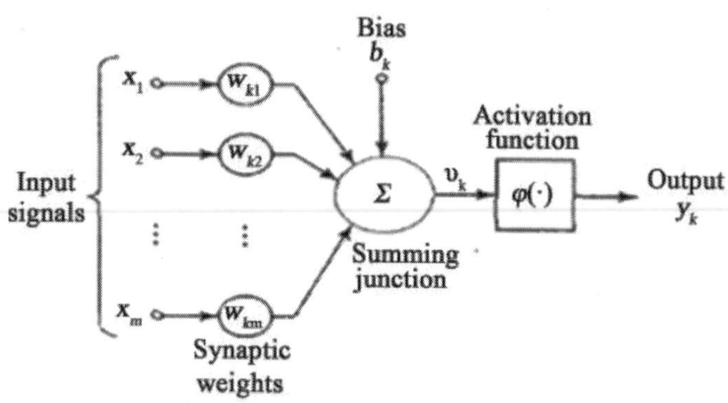

Figure 2. Models of a neuron.

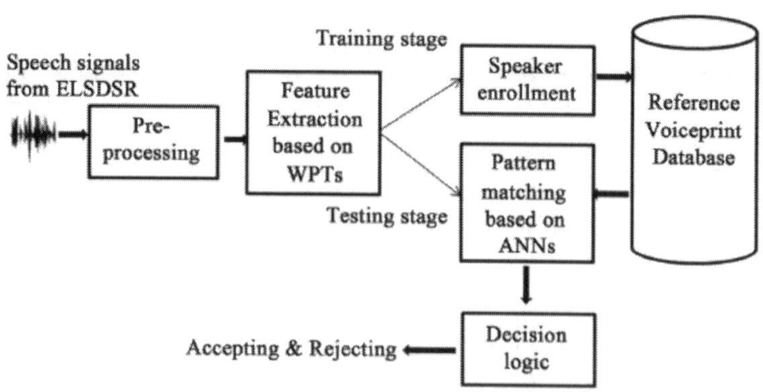

Figure 3. The proposed voice recognition system.

3.2. Pre-Processing of the Speech Signals

Signal pre-process have been carried out by performing normalization on the entered speech signals before feature extraction operation is carried to prevent the error estimation caused by speakers' volume changes. In other words, normalization makes the signals comparable regardless of differences in magnitude [2] . In this study, the signals are normalized by using the following equation [14] :

$$S_{pi} = \frac{Si - \mu}{\sigma}$$

where Si is the ith element of the signal S, μ and σ are the mean and standard deviation of the vector S, respectively; S_{Pi} is the ith element of the signal series S_P after normalization.

3.3. Feature Extraction by Wavelet Transforms

In FTs, the signal has been converted from the Time-domain into the Frequency-domain. In feature extraction we need to analysis the signal several times by using variable filters to extract a vital feature. As there are several languages, it is hard to choose a feature which carries a lot of information about each speaker but we investigated to extract a vital feature. This can be done by using STFT which contribute to calculate the MFCCs which is the familiar parameter in the speech recognition task. But MFCCs failed in the in the background noise as mentioned before. The wavelet transforms proved that it is an effective signal processing technique for a variety of signal processing problems. Here we decomposing the signal by using a suitable wavelet transform version and then calculate the energy enclosed in each signal as a distinguished feature for each speaker.

3.3.1. The Discrete Wavelet Transform

In Discrete Wavelet Transform (DWT), the signal is decomposed into Detail "D" and Approximation "A" coefficients in a dyadic form. This can be considered Multi-Resolution Analysis (MRA). In DWT, the scale and translation parameters are given by

$$a = a^j$$

$$b = kba^j$$

where k and j are integers. The function family becomes

$$\Psi_{j,k}(t) = a^{-j/2}\Psi\left(a^{-j}t - kb\right)$$

We can imagine that signals transformed by DWT as putting data into a series of high-pass and low-pass filters g(n), h(n) respectively. The high frequency content of speech signals which passed through the g(n) filter is called "details", and the low frequency content which passed through the h(n) filter is called "approximations", and only the approximation coefficients can be decomposed further as the decomposition level grows. The wavelet decomposition of the signal S analyzed at level j has the following structure [cAj, cD1, cDj]. DWT has been used at VRS but with limited success because of its left recursive nature [2] [7].

3.3.2. The Wavelet Packet Transform

In DWT, the high frequency components are removed, the voice will sound different but the speech can still be understood [2] . Wavelet packet analysis is an extension of the DWT [15] . Unlike the DWT, Wave PT decomposes both the high and low frequency bands at each iteration. A pair of low pass and high pass filters is used to generate two sub-bands with different frequencies and this considered one level of decomposition. These sub-bands are then down-sampled dyadically. This process can be repeated to partition the frequency spectrum into smaller frequency bands for resolving different features while localizing the temporal information. The complete binary tree for 3-level decomposition is produced and shown in the Figure 4.

In this study we have used WPT at level 7 with Daubechies 20-tap (db20), so we have divided the original signal into 128 sub-band signal. The Daubechies wavelet is a family of orthogonal wavelets, has a maximum number of vanishing moments, and conserves the energy of the signal and redistributed in a more compact form. The db20 is a member of the Daubechies family and is an estimation of the continuous form. It is asymmetric as shown in Figure 5.

As mentioned above due to conservation of energy the energy corresponding to each WP node is calculated according to the following equation

$$\sum_{n=-\infty}^{\infty} \left(\left|x(n)\right|\right)2$$

to form the feature vector of size(1 *128) which then used as input to the used Artificial Neural Network (ANN).

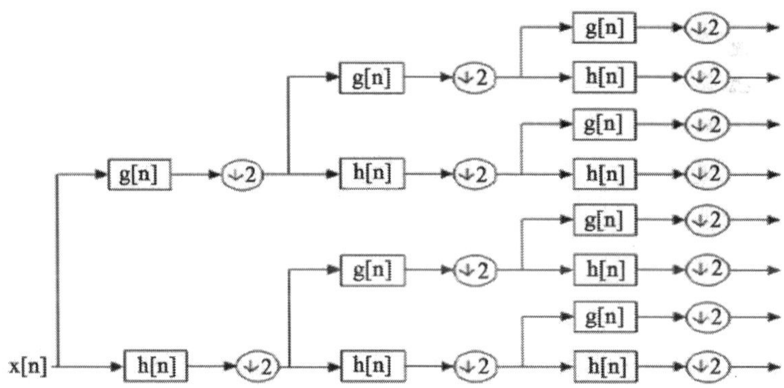

Figure 4. 3-level wavelet packet decomposition tree.

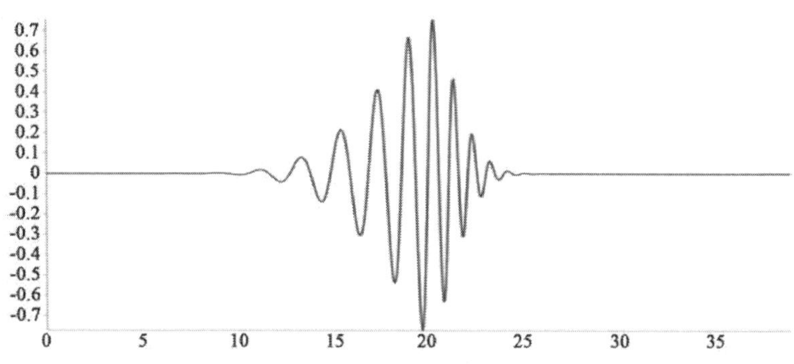

Figure 5. The db20 wavelet.

3.4. Classification by Artificial Neural Networks

In the design of the Speaker Recognition System (SRS), a classification technique of the features which extracted by DWT and WPT was used to evaluate the effectiveness of the extracted features in differentiating between the speakers. In this study, the classifier has been used was ANNs. ANNs are adaptive model that can changes its structure depended on information passes through it during the training process. Here the FFBPNN is used which belongs to the multilayer networks. After several experiments with various network structures the highest accuracy was found with 128 neurons for the two hidden layers, and 23 neurons in the output layer. The number of training epochs was selected to be 1000. The learning rate was left to be variable. The error Back- Propagation learning algorithm was used. The FFBPNN is proposed in Figure 6.

The recognition rate is evaluated by plotting the confusion matrix. It contains information about actual and predicted classifications done by a classification system.it consists of columns and rows where each column represents the predicted classes while each row represents the actual classes as proposed in Figure 7.

4. RESULTS AND DISCUSSION

A computer with specifications of processor Intel® Core™ 2 Duo CPU 2 GHZ, RAM 2 GB, an operating system windows 7 32-bit, and MATLAB (R2009b) program were used for achieving the experiment. A standard database ELSDSR is used where each registered speaker speaks a certain seven sentences and each sentence has been said for one time. Thus for each speaker we had seven speech samples, and 161 (23 *7) speech samples for the whole speakers. The experiment is carried out on three stages ranged from stage 0 to stage 2 and can be summarized as follows.

Stage 0, we had made normalization on each speech sample to prevent the error estimation caused by speakers' volume changes as see in Figure 8.

As noticed from Figure 8 the signal values are redistributed in a wide range vertically at this instance the speech samples are ready to analysis them to extract the features vectors and enter the next stage.

Stage 1, we have extracted the features vectors for each sentence for each speaker. In the features extraction step we have applied DWT and WPT but we have got a high performance rate when the speech samples were analyzed with WPT as it has characterized by the recursive analysis. We have used the Daubechies as mother wavelet and performed decomposition up to level 7 and the original speech sample is divided into 128 sub-band. We investigated on the better features that can represent each speaker, so calculation of the energy corresponding to each sub-band was the best choice as by it we can distinguish between the speakers' identities as proposed in Figure 9.

As noticed from the preceding figure the energy feature can differentiate between the speakers. We have formed a (128 *161) features matrix for the whole database's speakers. The features matrix was the threshold to enter the final stage in our experiment.

Stage 2, Classification is performed on two stages, in the training stage the features matrix and a suitable target matrix have been entered as inputsto the FFBPNN which specified in Table 1 and is drawn in Figure 6.

The FFBPNN varies its weight depended on the input and the expected output which feed to it. The expected output which also called the target matrix has a size of (23 *161). It is organized as for each speaker there is 7 columns (equals to the seven training sentences) with the first row has 1 values and the rest rows have 0 values, for the second speaker we bank the second seven columns with the second row in them has 1 values and the other rows have 0 values and so on until the twenty-three speaker for him the twenty-three row is filled with 1 values. Our target matrix can be summarized as in theTable 2.

Table 3 shows a summary of the confusion matrix which represents the accuracy of the FFBPNN for speakers individually as well as the system overall. Depending on the training database set, the system attempts to recognize 7 samples for each speaker where the total number of samples is 161. The overall system accuracy was 100%.

After getting on that accuracy in the training stage we saved the weights and biases of all layers of the trained FFBPNN. The weight values from the second hidden layer to the output layer can be abbreviated in Table 4.

At the testing stage, we have created another feature matrix (128 *23) where each speaker said another sentence differs from the seven sentences which used in the training stage and also differs from one speaker to another one, we have created the same FFBPNN with the saved weights and biases, and performed the test to get a recognition rate equal to 95.7% which recorded from the confusion matrix. Table 5 shows a summary of the confusion matrix in that stage. The system failed to recognize the six-teen speaker to become the overall recognized speakers equal to 22, and the overall accuracy of the system is 95.7%. Table 6 shows a comparison between MFCC, the Zak transform and the WPT techniques when all seven training files are used in the training phase. The WPT technique shows the same recognition rate 100% as the other two techniques if two test files are used and a recognition rate 95.7% if the first test file is used which better than the obtained by using the Zak transform 91.3% [5] .

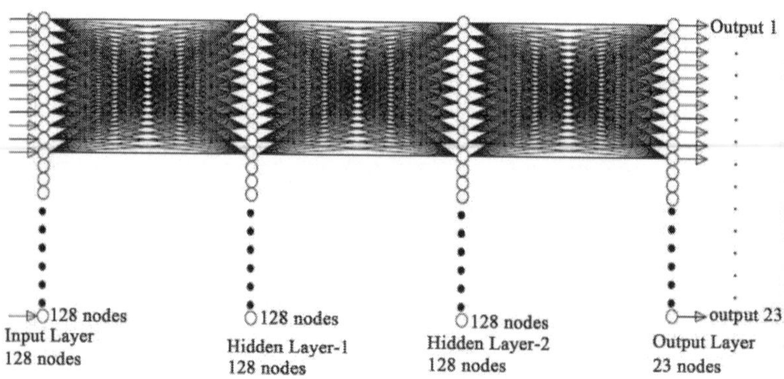

Figure 6. The FFBPNN structure.

		Predicted classes			
		Speaker #1	Speaker #2	Speaker #23	Accuracy%
Actual classes	Speaker #1	√			
	Speaker #2		√		
	⋮				
	Speaker #23			√	
	Total				Average%

Figure 7. The confusion matrix.

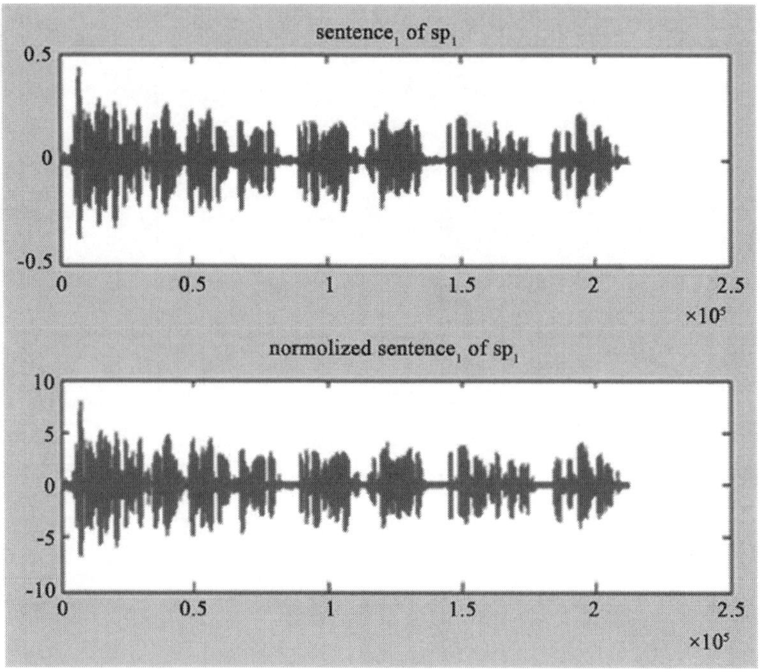

Figure 8. The speech signal for the first speaker before and after normalization.

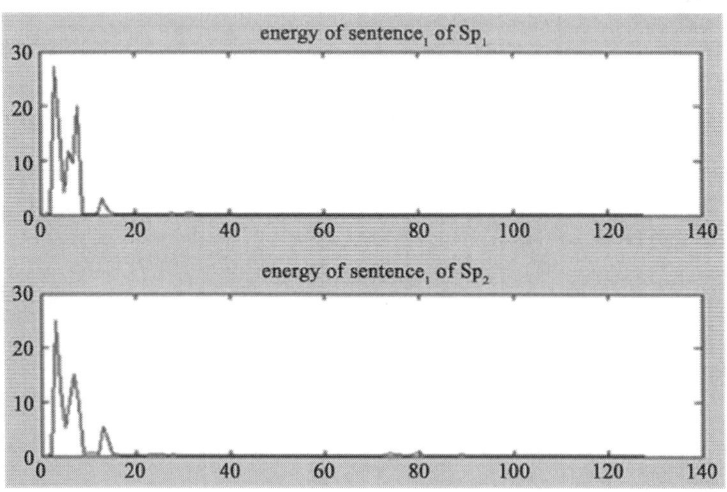

Figure 9. Energy distribution of sentence 1 for the two different speakers.

Table 1. Description of the FFBPNN.

Functions	Description
Network type	Feed forward back-propagation
Number of layers	Three layers: two hidden, one output
Number of neurons in each layer	128 neuron for each hidden, 23 neuron in the output, and 128 input node
Activation function	Tan-sigmoid
Training learning rule	Back-propagation
Number of epochs	1000
Learning rate	Varied

Table 2. The target matrix.

Speaker #1							Speaker #2…………………							Speaker #23						
1	1	1	1	1	1	1	0	0	0	0	0	0	0……………	0	0	0	0	0	0	0
0	0	0	0	0	0	0	1	1	1	1	1	1	1……………	0	0	0	0	0	0	0
0	0	0	0	0	0	0	0	0	0	0	0	0	0……………	0	0	0	0	0	0	0
0	0	0	0	0	0	0	0	0	0	0	0	0	0……………	0	0	0	0	0	0	0
0	0	0	0	0	0	0	0	0	0	0	0	0	0……………	0	0	0	0	0	0	0
⋮	⋮	⋮	⋮	⋮	⋮	⋮	⋮	⋮	⋮	⋮	⋮	⋮	⋮	⋮	⋮	⋮	⋮	⋮	⋮	⋮
0	0	0	0	0	0	0	0	0	0	0	0	0	0……………	1	1	1	1	1	1	1

Table 3. Recognition rate results at the Training stage.

Speaker	No. of signals	Recognized signals	Not recognized signals	Recognition rate %
Sp#1	7	7	0	100
Sp#2	7	7	0	100
Sp#3	7	7	0	100
Sp#4	7	7	0	100
Sp#5	7	7	0	100
Sp#6	7	7	0	100
Sp#7	7	7	0	100
Sp#8	7	7	0	100
Sp#9	7	7	0	100
Sp#10	7	7	0	100
Sp#11	7	7	0	100
Sp#12	7	7	0	100
Sp#13	7	7	0	100
Sp#14	7	7	0	100
Sp#15	7	7	0	100

Sp#15	7	7	0	100
Sp#16	7	7	0	100
Sp#17	7	7	0	100
Sp#18	7	7	0	100
Sp#19	7	7	0	100
Sp#20	7	7	0	100
Sp#21	7	7	0	100
Sp#22	7	7	0	100
Sp#23	7	7	0	100
Total	**161**	**161**	**0**	**100**

Table 4. Some of the weights values from the second hidden layer to the output layer.

Column #1	Column #2	Column #128
−0.09268249182061265	−0.15119278911501377	0.205933622253011566
−0.06370861164394098	−0.21319245828066766	0.022286562111564632
−0.1717592648717832	−0.14805606367740934	−0.12241513724860374
−0.0431523394284899	−0.18449418630223063	−0.13357997743075695
0.09903762471988278	−0.05582782956379674	0.20524026699714845
.	....	.
.	. ...	.
.	. ...	.
−0.139808274777698915	0.08471834362138755	0.1373001986105382

Table 5. Recognition rate results at the testing stage.

Speaker	No. of signals	Recognized signals	Not recognized signals	Recognition rate %
Sp#1	1	1	0	100
Sp#2	1	1	0	100
Sp#3	1	1	0	100
Sp#4	1	1	0	100
Sp#5	1	1	0	100
Sp#6	1	1	0	100
Sp#7	1	1	0	100
Sp#8	1	1	0	100
Sp#9	1	1	0	100
Sp#10	1	1	0	100
Sp#11	1	1	0	100
Sp#12	1	1	0	100
Sp#13	1	1	0	100
Sp#14	1	1	0	100
Sp#15	1	1	0	100
Sp#16	1	**0**	**1**	**0**

Sp#17	1	1	0	100
Sp#18	1	1	0	100
Sp#19	1	1	0	100
Sp#20	1	1	0	100
Sp#21	1	1	0	100
Sp#22	1	1	0	100
Sp#23	1	1	0	100
Total	**23**	**22**	**0**	**95.7**

Table 6. Comparison with MFCC, Zak in terms recognition rate.

Test Files	All Training Files	All Training Files	All Training Files
	MFFC	Zak	WPT
File 1	100%	91.30%	95.7%
Both	100%	100%	100%

5. CONCLUSION AND FUTURE WORK

Mohammed R. Saady, Hatem El-Borey, El-Sayed A. El-Dahshan, Ashraf Shamseldin Yahia (5,24)">5] on the same database where the recognition rate was 91.3% by using a single test file published before. In the future work, we will try to reduce the number of features associated with each speaker, and try to improve the efficiency of the proposed system in the real-time where the noise exist such as the street by using a hardware circuit such as the FPGA circuit.

REFERENCES

1. Desyatchikov, A.A., Kovkov, D.V., Lobantsov, V.V., Makovkin, K.A., Matveev, I.A., Murynin, A.B. and Chuchupal, V.Ya. (2006) A System of Algorithms for Stable Human Recognition. Journal of Computer and Systems Sciences International, 45, 958-969.

2. Wu, J.-D. and Lin, B.-F. (2009) Speaker Identification Using Discrete Wavelet Packet Transform Technique with Irregular Decomposition. Expert Systems with Applications, 36, 3136-3143.

3. Kinnunen, T. and Li, H.Z. (2010) An Overview of Text-Independent Speaker Recognition: From Features to Supervectors. Speech Communication, 52, 12-40.

4. Khalaf, E.F., Daqrouq, K. and Sherif, M. (2011) Wavelet Packet and Percent of Energy Distribution with Neural Networks Based Gender Identification System. Journal of Applied Sciences, 11, 2940-2946.

5. Hossen, A. and Al-Rawahi, S. (2010) A Text-Independent Speaker Identification System Based on the Zak Transform. Signal Processing: An International Journal, 4, 68-74.

6. ELSDSR Database for Speaker Recognition (2004).http://www.imm.dtu.dk/~lf/eLSDSR.htm

7. Lung, S.Y. (2006) Wavelet Feature Selection Based Neural Networks with Application to the Text Independent Speaker Identification. Pattern Recognition, 39, 1518-1521.

8. Avci, E. and Akpolat, Z.H. (2006) Speech Recognition Using a Wavelet Packet Adaptive Network Based Fuzzy Inference System. Expert Systems with Applications, 31, 495-503.

9. Sarikaya, R., Pellom, B.L. and Hansen, J.H.L. (1998) Wavelet Packet Transform Features with Application to Speaker Identification. Proceedings of the IEEE Nordic Signal Processing Symposium, Denmark, 81-84.

10. Goupillaud, P., Grossman, A. and Morlet, J. (1984) Cycle-Octave and Related Transforms in Seismic Signal Analysis. Geoexploration, 23, 85-102.

11. Louis, A.K., Maass, D. and Rieder, A. (1997) Wavelets-Theory and Applications. Wiley, Hoboken.

12. Avci, E. (2007) A New Optimum Feature Extraction and Classification Method for Speaker Recognition: GWPNN. Expert System with Applications, 32, 485-498.

13. Haykin, S. (1999) Neural Networks: A Comprehensive Foundation. 2nd Edition, Prentice-Hall, Englewood Cliffs.

14. Lou, X. and Loparo, K.A. (2004) Bearing Fault Diagnosis on Wavelet Transform and Fuzzy Inference. Mechanical System and Signal Processing, 18, 1077-1095.

15. Burrus, C.S., Gopinath, R.A. and Guo, H. (1998) Introduction to Wavelet and Wavelet Transforms. Prentice Hall, New Jersey.

Chapter 7

Cross-Word Arabic Pronunciation Variation Modeling Using Part of Speech Tagging

Dia AbuZeina, Husni Al-Muhtaseb and Moustafa Elshafei

[1] King Fahd University of Petroleum and Minerals, Dhahran, Saudi Arabia

1. INTRODUCTION

Speech recognition is often used as the front-end for many natural language processing (NLP) applications. Some of these applications include machine translation, information retrieval and extraction, voice dialing, call routing, speech synthesis/recognition, data entry, dictation, control, etc. Thus, much research work has been done to improve the speech recognition and the related NLP applications. However, speech recognition has some obstacles that should be considered. Pronunciation variations and small words misrecognition are two major problems that lead to performance reduction. Pronunciation variations problem can be divided into two parts: within-word variations and cross-word variations. These two types of pronunciation variations have been tackled by many researchers using different approaches. For example, cross-word problem can be solved using phonological rules and/or small-word merging. (AbuZeina et al., 2011a) used the phonological rules to model cross-word variations for Arabic. For English, (Saon & Padmanabhan, 2001) demonstrated that short words are more frequently misrecognized, they also had achieved a statistically significant enhancement using small-word merging approach.

An automatic speech recognition (ASR) system uses a decoder to perform the actual recognition task. The decoder finds the most likely words sequence for the given utterance using Viterbi algorithm. The ASR decoder task might be seen as an alignment process between the observed phonemes and the reference phonemes (dictionary phonemic transcription). Intuitively, to have a better accuracy in any alignment process, long sequences are highly favorable instead of short ones. As such, we expect enhancement if we merge words (short or long). Hence fore, a thorough investigation was performed on Arabic speech to discover a suitable merging cases. We found that Arabic speakers usually augment two consecutive words; a noun that is followed by an adjective and a preposition that is followed by a word. Even though we believe that other cases are found in Arabic speech, we chose two cases to validate our proposed method. Among the ASR components, the pronunciation dictionary and the language model were used to model our above mentioned objective. This means that the acoustic models for the baseline and the enhanced method are the same.

This research work is conducted for Modern Standard Arabic (MSA). So, the work will necessarily contain many examples in Arabic. Therefore, it would be appropriate for the reader if we start first by providing a Romanization (Ryding, 2005) of the Arabic letters and diacritical marks. Table 1 shows the Arabic–Roman letters mapping table. The diacritics Fatha, Damma, and Kasra are represented using a, u, and i, respectively.

To validate the proposed method, we used Carnegie Mellon University (CMU) Sphinx speech recognition engine. Our baseline system contains a pronunciation dictionary of 14,234 words from a 5.4 hours pronunciation corpus of MSA broadcast news. For tagging, we used the Arabic module of Stanford tagger. Our results show that part of speech (PoS) tagging is considered a promising track to enhance Arabic speech recognition systems.

The rest of this chapter is organized as follows. Section 2 presents the problem statement. Section 3 demonstrates the speech recognition components. In Section 4, we differentiate between within-word and cross-word pronunciation variations followed by the Arabic speech recognition in Section 5. The proposed method is presented in Section 6 and the results in Section 7. The discussion is provided in Section 8. In Section 9, we highlight some of the future directions. We conclude the work in Section 10.

Table 1. Arabic–Roman letters mapping table

Arabic	Roman	Arabic	Roman	Arabic	Roman	Arabic	Roman
ء (hamza)	’	د (daal)	d	ض (Daad)	D	ك (kaaf)	k
ب (baa’)	b	ذ (dhaal)	dh	ط (Taa’)	T	ل (laam)	l
ت (taa’)	t	ر (raa’)	r	ظ (Zaa’)	Z	م (miim)	m
ث (thaa’)	th	ز (zaay)	z	ع (‘ayn)	‘	ن (nuun)	n
ج (jiim)	j	س (siin)	s	غ (ghayn)	gh	ه (haa’)	h
ح (Haa’)	H	ش (shiin)	sh	ف (faa’)	f	و (waaw)	w or u
خ (khaa’)	kh	ص (Saad)	S	ق (qaaf)	q	ي (yaa’)	y or ii

2. PROBLEM STATEMENT

Continuous speech is characterized by augmenting adjacent words, which do not occur in isolated speech. Therefore, handling this phenomenon is a major requirement in continuous speech recognition systems. Even though Hidden Markov Models (HMMs) based ASR decoder uses triphones to alleviate the negative effects of cross-word phenomenon, more effort is still needed to model some cross-word cases that could not be avoided using triphones. In continuous ASR systems, the dictionary is usually initiated using corpus transcription words, i.e. each word is considered as an independent entity. In this case, speech cross-word merging will reduce the performance. Two main methods are usually used to model the cross-word problem, phonological rules and small-word merging. Even though the phonological rules and small-word merging methods enhance the performance, we believe that generating compound words is also possible using PoS tagging.

Initially, there are two reasons why cross-word modeling is an effective method in speech recognition system: First, the speech recognition problem appears as an alignment process, hence for, having long sequences is better than short ones as demonstrated by (Saon and Padmanabhan, 2001). To illustrate the effect of co-articulation phenomenon (merging of words in continuous speech), let us examineFigure 1 and Figure 2. Figure 1 shows the words to be considered with no compound words, whileFigure 2 shows the words with compound words. In both figures we represented the hypotheses words using bold black lines. During decoding, the ASR decoder will investigate many words and hypotheses. Intuitively, the ASR decoder will choose the long words instead of two short words. The difference between the two figures is the total number of words that will be considered during the decoding process. Figure 2 shows that the total number of words for the hypotheses is less than the total words in Figure 1 (Figure 1 contains 34 words while Figure 2 contains 18 words). Having less number of total words during decoding process means having less decoding options (i.e. less ambiguity), which is expected to enhance the performance.

Second, compounding words will lead to more robust language model. the compound words which are represented in the language model will provide better representations of words relations. Therefore, enhancement is expected as correct choice of a word will increase the probability of choosing a correct neighbor words. The effect of compounding words was investigated by (Saon & Padmanabhan, 2001). They mathematically demonstrated that compound words enhance the language model performance, therefore, enhancing the overall recognition output. They showed that the compound words have the effect of incorporating a trigram dependency in a bigram language model. In general, the compound words are most likely to be correctly recognized more than two separated words. Consequently, correct recognition of a word might lead to another correct word through the enhanced N-grams language model. In contrast, misrecognition of a word may lead to another misrecognition in the adjacent words and so on.

For more clarification, we present some cases to show the short word misrecognition, and how is the long word is much likely to be recognized correctly. Table 2 shows three speech files that were tested in the baseline and the enhanced system. Of course, it is early to show some results, but we see that it is worthy to support our motivation claim. In Table 2, it is clear that the misrecognitions were mainly occurred in the short words (the highlighted short words were misrecognized in the baseline system).

In this chapter, the most noticeable Arabic ASRs performance reduction factor, the cross-word pronunciation variations, is investigated. To enhance speech recognition accuracy, a knowledge-based technique was utilized to model the cross-word pronunciation variation at two ASR components: the pronunciation dictionary and the language model. The proposed knowledge-based approach method utilizes the PoS tagging to compound consecutive words according to their tags. We investigated two pronunciation cases, a noun that is followed by an adjective, and a preposition that is followed by a word. the proposed method showed a significant enhancement.

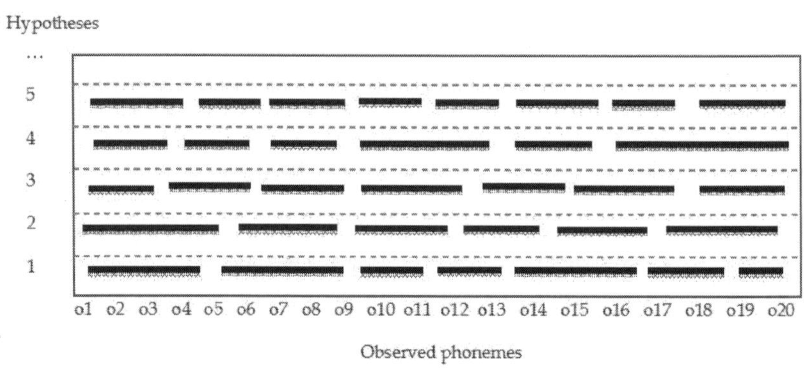

Figure 1. A list of hypotheses without compounding words

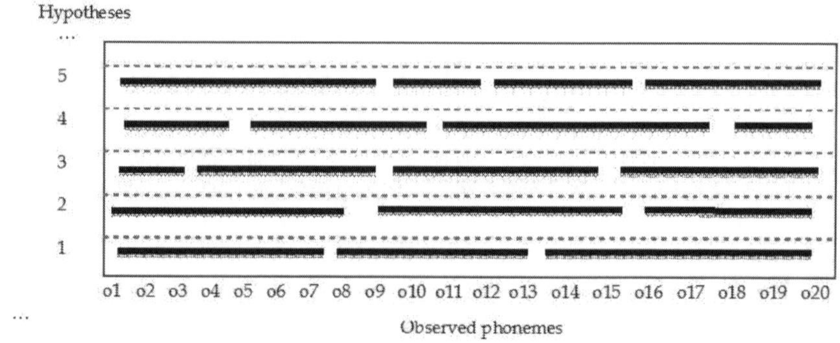

Figure 2. A list of hypotheses with compounding words

3. SPEECH RECOGNITION

Modern large vocabulary, speaker-independent, continuous speech recognition systems have three knowledge sources, also called linguistic databases: acoustic models, language model, and pronunciation dictionary (also called lexicon). Acoustic models are the HMMs of the phonemes and triphones (Hwang, 1993). The language model is the module that provides the statistical representations of the words sequences based on the transcription of the text corpus. The dictionary is the module that serves as an intermediary between the acoustic model and the language model. The dictionary contains the words available in the language and the pronunciation of each word in terms of the phonemes available in the acoustic models.

Figure 3 illustrates the sub-systems that are usually found in a typical ASR system. In addition to the knowledge sources, an ASR system contains a Front-End module which is used to convert the input sound into feature vectors to be usable by the rest of the system. Speech recognition systems usually use feature vectors that are based on Mel Frequency Cepstral Coefficients (MFCCs), (Rabiner and Juang, 2004).

The following is a brief introduction to typical ASR system components. The reader can find more elaborate discussion in (Jurafsky and Martin, 2009).

Table 2. Illustrative cross-word misrecognition results

The speech files to be tested	سَيَتَقَابَلانِ وَجهًا لِوَجه فِي المُبَارَاةِ النَّهَائِيَّة sayataqabalani wajhan liwajh fy ’lmubarah ’lniha’iya وَمُمَثِّلِينَ عَن عَدَدٍ مِن الدُّوَلِ الأُورُوبِيَّة wamumathilyna ‘an ‘adadin mina ’lduwali ’l’wrubiya
The baseline system results	سَيَتَقَابَلانِ وَجهًا لِوَجه المُبَارَاةِ النَّهَائِيَّة sayataqabalani wajhan liwajh ’lmubarah ’lniha’iya وَمُمَثِّلِين عَن إِنَّ الدُّوَلِ الأُورُوبِيَّة wamumathilyna ‘an ’inna ’lduwali ’l’wrubiya
The enhance system results	سَيَتَقَابَلانِ وَجهًا لِوَجه فِي المُبَارَاةِ النَّهَائِيَّة sayataqabalani wajhan liwajh fy ’lmubarah ’lniha’iya وَمُمَثِّلِينَ عَن عَدَدٍ مِن الدُّوَلِ الأُورُوبِيَّة wamumathilyna ‘an ‘adadin mina ’lduwali ’l’wrubiya

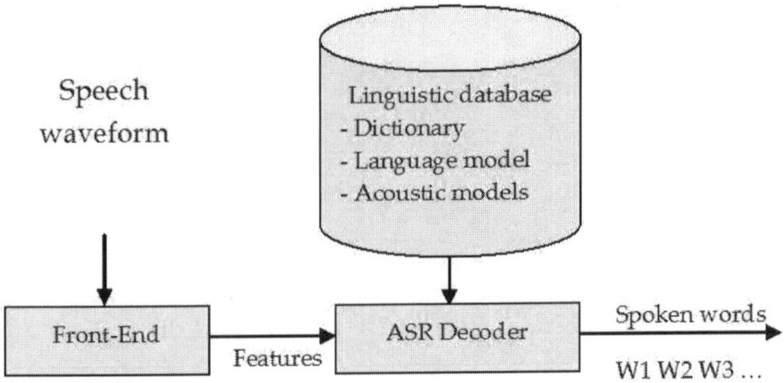

Figure 3. An ASR architecture

3.1. Front-End

The purpose of this sub-system is to extract speech features which play a crucial role in speech recognition performance. Speech features includes Linear Predictive Cepstral Coefficients (LPCC), MFCCs and Perceptual Linear Predictive (PLP) coefficients. The Sphinx engine used in this work is based on MFCCs.

The feature extraction stage aims to produce the spectral properties (features vectors) of speech signals. The feature vector consists of 39 coefficients. A speech signal is divided into overlapping short segments that are represented using MFCCs. Figure 4 shows the steps to extract the MFCCs of a speech signal (Rabiner & Juang, 2004). These steps are summarized below.

Sampling and Quantization: Sampling and quantization are the two steps for analog-to-digital conversion. The sampling rate is the number of samples taken per second, the sampling rate used in this study is 16 k samples per seconds. The quantization is the process of representing real-valued numbers as integers. The analysis window is about 25.6 msec (410 samples), and consecutive frames overlap by 10 msec.

Preemphasis: This stage is to boost the high frequency part that was suppressed during the sound production mechanism, so making the information more available to the acoustic model.

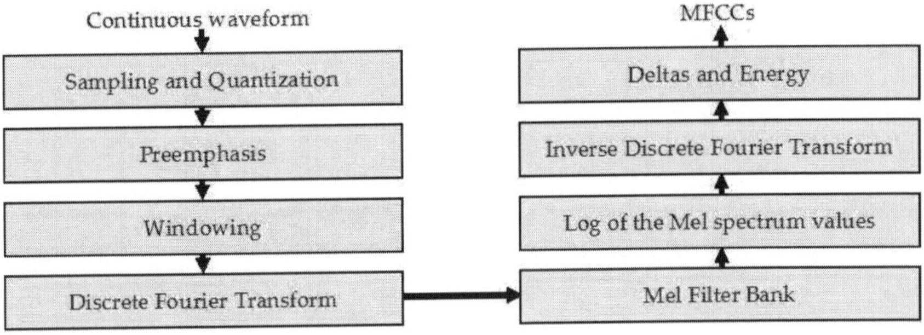

Figure 4. Feature vectors extraction

Windowing: Each analysis window is multiplied by a Hamming window.

Discrete Fourier Transform: The goal of this step is to obtain the magnitude frequency response of each frame. The output is a complex number representing the magnitude and phase of the frequency component in the original signal.

Mel Filter Bank: A set of triangular filter banks is used to approximate the frequency resolution of the human ear. The Mel frequency scale is linear up to 1000 Hz and logarithmic thereafter. For 16 KHz sampling rate, Sphinx engine uses a set of 40 Mel filters.

Log of the Mel spectrum values: The range of the values generated by the Mel filter bank is reduced by replacing each value by its natural logarithm. This is done to make the statistical distribution of spectrum approximately Gaussian.

Inverse Discrete Fourier Transform: This transform is used to compress the spectral information into a set of low order coefficients which is called the Mel-cepstrum. Thirteen MFCC coefficients are used as a basic feature vector,

$xt(k) \quad 0 \leq k \leq 12$

Deltas and Energy: For continuous models, the 13 MFCC parameters along with computed delta and delta-deltas parameters are used as a single stream 39 parameters feature vector. For semi-continuous models, x(0) represents the log Mel spectrum energy, and is used separately to derive other feature parameters, in addition to the delta and double delta parameters. Figure 5 shows part of the feature vector of a speech file after completing the feature extraction process. Each column represents the basic 13 features of a 25.6 milliseconds frame.

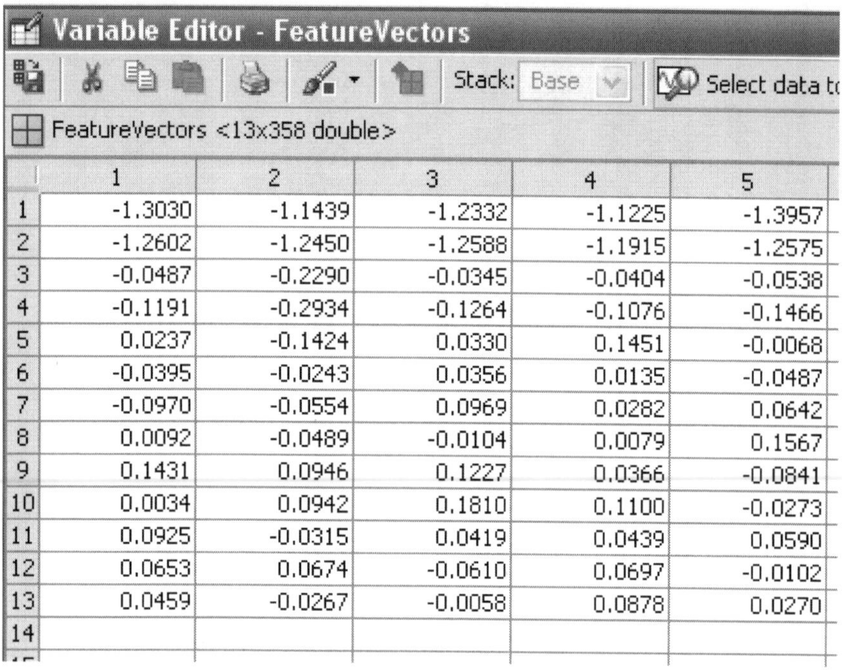

	1	2	3	4	5
1	-1.3030	-1.1439	-1.2332	-1.1225	-1.3957
2	-1.2602	-1.2450	-1.2588	-1.1915	-1.2575
3	-0.0487	-0.2290	-0.0345	-0.0404	-0.0538
4	-0.1191	-0.2934	-0.1264	-0.1076	-0.1466
5	0.0237	-0.1424	0.0330	0.1451	-0.0068
6	-0.0395	-0.0243	0.0356	0.0135	-0.0487
7	-0.0970	-0.0554	0.0969	0.0282	0.0642
8	0.0092	-0.0489	-0.0104	0.0079	0.1567
9	0.1431	0.0946	0.1227	0.0366	-0.0841
10	0.0034	0.0942	0.1810	0.1100	-0.0273
11	0.0925	-0.0315	0.0419	0.0439	0.0590
12	0.0653	0.0674	-0.0610	0.0697	-0.0102
13	0.0459	-0.0267	-0.0058	0.0878	0.0270
14					

Figure 5. Snapshot of the MFCCs of a speech file

3.2. Linguistic Database

This part contains the modifications required for a particular language. It contains three parts: acoustic models, language model, and pronunciation dictionary. Acoustic models contain the HMMs used in recognition process. The language model contains language's words and their combinations, each combination has two or three words. A pronunciation dictionary contains the words and their pronunciation phonemes.

3.2.1. Acoustic Models

Acoustic models are statistical representations of the speech phones. Precise acoustic model is a key factor to improve recognition accuracy as it characterizes the HMMs of each phone. Sphinx uses 39 English phonemes (The CMU Pronunciation Dictionary, 2011). The acoustic models use a 3- to 5-state Markov chain to represent the speech phone (Lee, 1988). Figure 6 shows a representation of a 3-state phone's acoustic model. In Figure 6, S1 is the representation of phone at the beginning, while S2 and S3 represent of the phone at the middle and the end states, respectively. Associated with S1, S2, and S3 are state emission probabilities, $b_j(x_t) = P(o = x_t \mid S_t = j)$, representing the probability of observing the feature vector in the state j. The emission probabilities are usually modeled by Gaussian mixture densities.

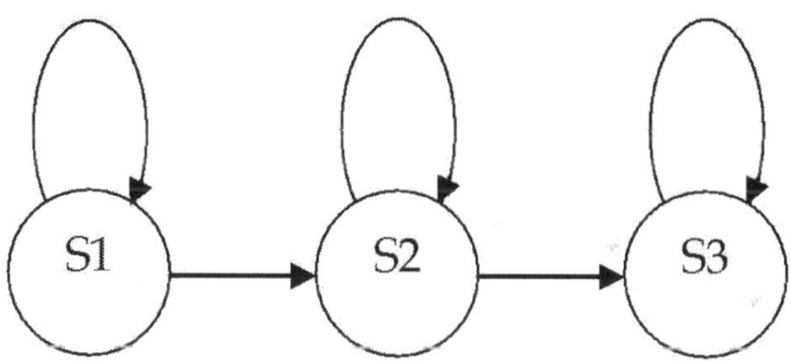

Figure 6. State phone acoustic model

In continuous speech, each phoneme is influenced in different degrees by its neighboring phonemes. Therefore, for better acoustic modeling, Sphinx uses triphones. Triphones are context dependent models of phonemes; each triphone represents a phoneme surrounded by specific left and right phonemes (Hwang, 1993).

3.2.2. Language Model

The N-gram language model is trained by counting N-gram occurrences in a large transcription corpus to be then smoothed and normalized. In general, an N-gram language model is used to calculate the probability of a given sequence of words as follows:

$$P(w_1^n) = \prod_{k=1}^{n} p(w_k | w_1^{k-1})$$

Where n is limited to include the words' history as bigram (two consequent words), trigram (three consequent words), 4-gram (four consequent words), etc. for example, by assigning n=2, the probability of a three word sequence using bigram is calculated as follows:

$$P(w_1 w_2 w_3) = p(w_3 | w_2) p(w_2 | w_1) p(w_1)$$

The CMU statistical language tool is described in (Clarkson & Rosenfeld, 1997). The CMU statistical language tool kit has been used to generate our Arabic statistical language model. Figure 7 shows the steps for creation and testing the language model, the steps are:

- Compute the word unigram counts.
- Convert the word unigram counts into a vocabulary list.
- Generate bigram and trigram tables based on this vocabulary.

The CMU language modeling tool comes with a tool for evaluating the language model. The evaluation measures the perplexity as indication of the convenient (goodness) of the language model. For more information of the perplexity, please refer to Section 7.

3.2.3. Pronunciation Dictionary

Both training and recognition stages require a pronunciation dictionary which is a mapping table that maps words into sequences of phonemes. A pronunciation dictionary is basically designed to be used with a particular set of words. It provides the pronunciation of the vocabulary for the transcription corpus using the defined phoneme set. Like acoustic models and language model, the performance of a speech recognition system depends critically on the dictionary and the phoneme set used to build the dictionary. In decoding stage, the dictionary serves as intermediary between the acoustic model and the language model.

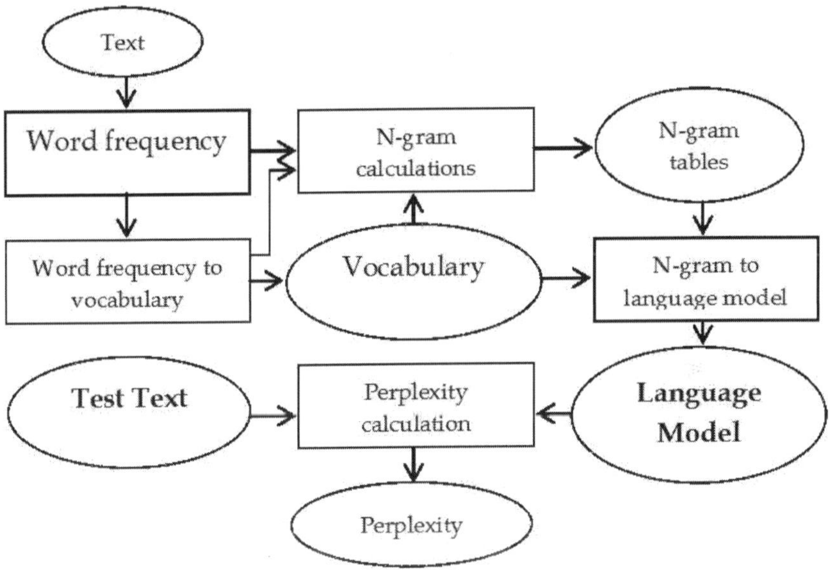

Figure 7. Steps for creating and testing language model

There are two types of dictionaries: closed vocabulary dictionary and open vocabulary dictionary. In closed vocabulary dictionary, all corpus transcription words are listed in the dictionary. In contrast, it is possible to have non-corpus transcription words in the open vocabulary dictionary. Typically, the phoneme set, that is used to represent dictionary words, is manually designed by language experts. However, when human expertise is not available, the phoneme set is possible to be selected using data-driven approach as demonstrated by (Singh et al. 2002). In addition to providing phonemic transcriptions of the words of the target vocabulary, the dictionary is the place where alternative pronunciation variants are added such as in (Ali et al., 2009) for Arabic.

3.3. Decoder (Recognizer)

With help from the linguistic part, the decoder is the module where the recognition process takes place. The decoder uses the speech features presented by the Front-End to search for the most probable words and, then, sentences that correspond to the observed speech features. The recognition process starts by finding the likelihood of a given sequence of speech features based on the phonemes HMMs.

The speech recognition problem is to transcribe the most likely spoken words given the acoustic observations. If O=o1,o,2....on is the acoustic observation, and W=w1,w,2....wn is a word sequence, then:

$$\widehat{W} = \underbrace{arg\ max}_{for\ all\ words}\ P(W)P(O|W)$$

Where WW is the most probable word sequence of the spoken words, which is also called maximum posteriori probability. P(W) is the prior probability computed in the language model, and P(O|W) is the probability of observation computed using the acoustic model.

4. PRONUNCIATION VARIATION

The main goal of ASRs is to enable people to communicate more naturally and effectively. But this ultimate dream faces many obstacles such as different speaking styles which lead to "pronunciation variation" phenomenon. This phenomenon appears in the form of insertions, deletions, or substitutions of phoneme(s) relative to the phonemic transcription in the pronunciation dictionary. (Benzeghiba et al., 2007) presented the speech variability sources: foreign and regional accents, speaker physiology, spontaneous speech, rate of speech, children speech, emotional state, noises, new words, and more. Accordingly, handling these obstacles is a major requirement to have better ASR performance.

There are two types of pronunciation variations: cross-word variations and within-word variations. A within-word variation causes alternative pronunciation(s) of the same word. In contrast, a cross-word variation occurs in continuous speech in which a sequence of words forms a compound word that should be treated as a one entity. The pronunciation variation can be modeled in two approaches: knowledge-based and data-driven. Knowledge-based depends on linguistic studies that lead to the phonological rules which are called to find the possible alternative variants. On the other hand, data-driven methods depend solely on the pronunciation corpus to find the pronunciation variants (direct data-driven) or transformation rules (indirect data-driven). In this chapter, we will use the knowledge-based approach to model the cross-word pronunciation variation problem.

As pros and cons of both approaches, the knowledge-based approach is not exhaustive; not all of the variations that occur in continuous speech have been described. Whereas obtaining reliable information using data-driven is difficult. However, (Amdal & Fossler-Lussier 2003) mentioned that there is a growing interest in data-driven methods over the knowledge-based methods due to lack of domains expertise. Figure 8 displays these two techniques. Figure 8 also distinguishes between the types of variations and the modeling techniques by a dashed line. The pronunciation variation types are above the dashed line whereas the modeling techniques are under the dashed line.

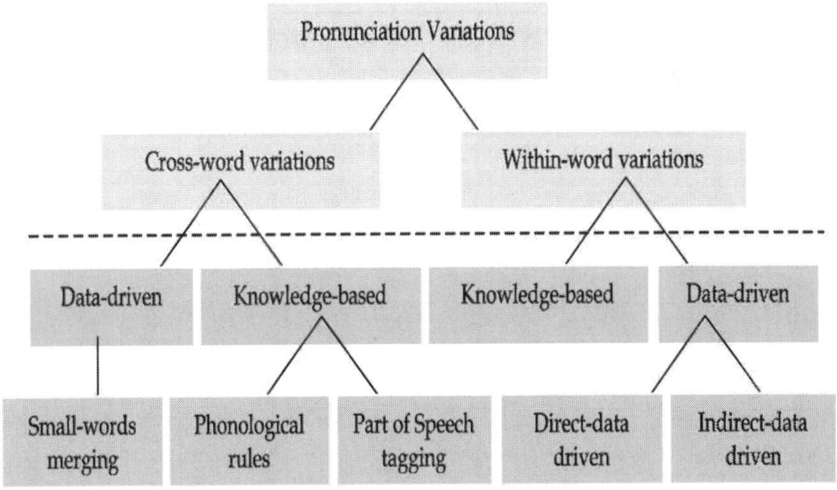

Figure 8. Pronunciation variations and modeling techniques

5. ARABIC SPEECH RECOGNITION

This work focuses on Arabic speech recognition, which has gained increasing importance in the last few years. Arabic is a Semitic language spoken by more than 330 million people as a native language (Farghaly & Shaalan, 2009). While Arabic language has many spoken dialects, it has a standard written language. As a result, more challenges are introduced to speech recognition systems as the spoken dialects are not officially written. The same country could contain different dialects and a dialect itself can vary from region to another according to different factors such as religion, gender, urban/rural, etc. Speakers with different dialects usually use modern standard Arabic (MSA) to communicate.

5.1. Modern Standard Arabic

In this chapter, we consider the modern standard Arabic (MSA) which is currently used in writing and in most formal speech. MSA is also the major medium of communication for public speaking and news broadcasting (Ryding, 2005) and is considered to be the official language in most Arabic-speaking countries (Lamel et al., 2009). Arabic language challenges will be presented in the next section. Followed by the literature review and recent results efforts in Arabic speech recognition. For more information about modern standard Arabic, (Ryding, 2005) is a rich reference.

5.2. Arabic Speech Recognition Challenges

Arabic speech recognition faces many challenges. First, Arabic has many dialects where same words are pronounced differently. In addition, the spoken dialects are not officially written, it is very costly to obtain adequate corpora, which present a training problem for the Arabic ASR researchers (Owen et al., 2006). Second, Arabic has short vowels (diacritics), which are usually ignored in text. The lack of diacritical marks introduces another serious problem to Arabic speech recognition. Consequently, more hypotheses' words will be considered during decoding process which may reduce the accuracy. (Elmahdy et al., 2009) summarized some of the problems raised in Arabic speech recognition. They highlighted the following problems: Arabic phonetics, diacritization problem, grapheme-to-phoneme, and morphological complexity. Although foreign phoneme sounds as /v/ and /p/ are used in Arabic speech in foreign names, the standard Arabic letters do not have standard letter assigned for foreign sounds. Second, the absence of the diacritical marks in modern Arabic text creates ambiguities for pronunciations and meanings. For example, the non-diacritized Arabic word (كتب) could be read as one of several choices, some of which are: (كَتَبَ,he wrote), (كُتِبَ, it was written), and (كُتُب, books). Even though, an Arabic reader can interpret and utter the correct choice, it is hard to embed this cognitive process in current speech recognition and speech synthesis systems. The majority of Arabic corpora available for the task of acoustic modeling have non-diacritized transcription. (Elmahdy et al., 2009) also showed that grapheme-to-phoneme relation is only true for diacritized Arabic script. Hence fore, Arabic speech recognition has an obstacle because the lack of diacritized corpora. Arabic morphological complexity is demonstrated by the large number of affixes (prefixes, infixes, and suffixes) that can be added to the three consonant radicals to form patterns. (Farghaly& Shaalan, 2009) provided a comprehensive study of Arabic language challenges and solutions. The mentioned challenges include: the nonconcatenative nature of Arabic morphology, the absence of the orthographic representation of Arabic diacritics from contemporary Arabic text, and the need for an explicit grammar of MSA that defines linguistic constituency in the absence of case marking. (Lamel et al., 2009) presented a number of challenges for Arabic speech recognition such as no diacritics, dialectal variants, and very large lexical variety. (Alotaibi et al., 2008) introduced foreign-accented Arabic speech as a challenging task in speech recognition. (Billa et al., 2002) discussed a number of research issues for Arabic speech recognition, e.g., absence of diacritics in written text and the presence of compound words that are formed by the concatenation of certain conjunctions, prepositions, articles, and pronouns, as prefixes and suffixes to the word stem.

5.3. Literature and Recent Work

A number of researchers have recently addressed development of Arabic speech recognition systems. (Abushariah et al., 2012) proposed a framework for the design and development of a speaker-independent continuous automatic Arabic speech recognition system based on a phonetically rich and balanced speech corpus. Their method reduced the WER to 9.81% for a diacritized transcription corpus, as they have reported. (Hyassat & Abu Zitar, 2008) described an Arabic speech recognition system based on Sphinx 4. Three corpora were developed, namely, the Holy

Qura'an corpus of about 18.5 hours, the command and control corpus of about 1.5 hours, and the Arabic digits corpus of less than 1 hour of speech. They also proposed an automatic toolkit for building pronunciation dictionaries for the Holy Qur'an and standard Arabic language. (Al-Otaibi, 2001)] provided a single-speaker speech dataset for MSA. He proposed a technique for labeling Arabic speech. using the Hidden Markov Model Toolkit (HTK), he reported a recognition rate for speaker dependent ASR of 93.78%. (Afify et al., 2005) compared grapheme-based recognition system with explicitly modeling diacritics (short vowels). They found that a diacritic modeling improves recognition performance. (Satori et al., 2007) used CMU Sphinx tools for Arabic speech recognition. They demonstrated the use of the tools for recognition of isolated Arabic digits. They achieved a digits recognition accuracy of 86.66% for data recorded from six speakers. (Alghamdi et al., 2009) developed an Arabic broadcast news transcription system. They used a corpus of 7.0 h for training and 0.5 h for testing. The WER they obtained was 14.9%. (Lamel et al., 2009) described the incremental improvements to a system for the automatic transcription of broadcast data in Arabic, highlighting techniques developed to deal with specificities (no diacritics, dialectal variants, and lexical variety) of the Arabic language. (Billa et al., 2002) described the development of audio indexing system for broadcast news in Arabic. Key issues addressed in their work revolve around the three major components of the audio indexing system: automatic speech recognition, speaker identification, and named entity identification. (Soltau et al., 2007) reported advancements in the IBM system for Arabic speech recognition as part of the continuous effort for the Global Autonomous Language Exploitation (GALE) project. The system consisted of multiple stages that incorporate both diacritized and non-diacritized Arabic speech model. The system also incorporated a training corpus of 1,800 hours of unsupervised Arabic speech. (Azmi et al., 2008) investigated using Arabic syllables for speaker-independent speech recognition system for Arabic spoken digits. The pronunciation corpus used for both training and testing consisted of 44 Egyptian speakers. In a clean environment, experiments showed that the recognition rate obtained using syllables outperformed the rate obtained using monophones, triphones, and words by 2.68%, 1.19%, and 1.79%, respectively. Also in noisy telephone channel, syllables outperformed the rate obtained using monophones, triphones, and words by 2.09%, 1.5%, and 0.9%, respectively. (Elmahdy et al., 2009) used acoustic models trained with large MSA news broadcast speech corpus to work as multilingual or multi-accent models to decode colloquial Arabic. (Khasawneh et al., 2004) compared the polynomial classifier that was applied to isolated-word speaker-independent Arabic speech and dynamic time warping (DTW) recognizer.

They concluded that the polynomial classifier produced better recognition performance and much faster testing response than the DTW recognizer. (Shoaib et al., 2003) presented an approach to develop a robust Arabic speech recognition system based on a hybrid set of speech features. The hybrid set consisted of intensity contours and formant frequencies. (Alotaibi, 2004) reported achieving high-performance Arabic digits recognition using recurrent networks. (Choi et al., 2008) presented recent improvements to their English/Iraqi Arabic speech-to-speech translation system. The presented system-wide improvements included user interface, dialog manager, ASR, and machine translation components. (Nofal et al., 2004) demonstrated a design and implementation of stochastic-based new acoustic models for use with a command and control system speech recognition system for the Arabic. (Mokhtar & El-Abddin, 1996) represented the techniques and algorithms used to model the acoustic-phonetic structure of Arabic speech recognition using HMMs. (Park et al., 2009) explored the training and adaptation of multilayer perceptron (MLP) features in Arabic ASRs. They used MLP features to incorporate short-vowel information into the graphemic system. They also used linear input networks (LIN) adaptation as an alternative to the usual HMM-based linear adaptation. (Imai et al.,1995) presented a new method for automatic generation of speaker-dependent phonological rules in order to decrease recognition errors caused by pronunciation variability dependent on speakers. (Muhammad et al., 2011) evaluated conventional ASR system for six different types of voice disorder patients speaking Arabic digits. MFCC and Gaussian mixture models (GMM)/HMM were used as features and classifier, respectively. Recognition result was analyzed for recognition for types of diseases. (Bourouba et al., 2006) presented a HMM/support vectors machine (SVM) (k-nearest neighbor) for recognition of isolated spoken Arabic words. (Sagheer et al., 2005) presented a visual speech features representation system. They used it to comprise a complete lip-reading system. (Taha et al., 2007) demonstrated an agent-based design for Arabic speech recognition. They defined the Arabic speech recognition as a multi-agent system where each agent had a specific goal and deals with that goal only. (Elmisery et al., 2003) implemented a pattern matching algorithm based on HMM using field programmable gate array (FPGA). The proposed approach was used for isolated Arabic word recognition. (Gales et al., 2007) described the development of a phonetic system for Arabic speech recognition. (Bahi & Sellami, 2001) presented experiments performed to recognize isolated Arabic words. Their recognition system was based on a combination of the vector quantization technique at the acoustic level and markovian modeling. (Essa et al., 2008) proposed a combined classifier architectures based on Neural Networks by varying the initial weights, architecture, type, and training data to recognize Arabic isolated words. (Emami & Mangu, 2007) studied the use of neural network language models (NNLMs) for Arabic broadcast news and broadcast conversations speech recognition. (Messaoudi et al., 2006) demonstrated that by building a very large vocalized vocabulary and by using a language model including a vocalized component, the WER could be significantly reduced. (Vergyri et al., 2004) showed that the use of morphology-based language models at different stages in a large vocabulary continuous speech recognition (LVCSR) system for Arabic leads to WER reductions. To deal with the huge lexical variety, (Xiang et al., 2006) concentrated on the transcription of Arabic broadcast news by utilizing morphological decomposition in both acoustic and language modeling in their system. (Selouani & Alotaibi, 2011) presented genetic algorithms to adapt HMMs for non-native speech in a large vocabulary speech recognition system of MSA. (Saon et al., 2010) described the Arabic broadcast transcription system fielded by IBM in the GALE project. they reported improved discriminative training, the use of subspace Gaussian mixture models (SGMM), the use of neural network acoustic features, variable frame rate decoding, training data partitioning experiments, unpruned n-gram language models, and neural network based language modeling (NNLMs). The achieved WER was 8.9% on the evaluation test set. (Kuo et al., 2010) studied various syntactic and morphological context features incorporated in an NNLM for Arabic speech recognition.

6. THE PROPOSED METHOD

Since the ASR decoder works better with long words, our method focuses on finding a way to merge transcription words to increase the number of long words. For this purpose, we consider to merge words according to their tags. That is, merge a noun that is followed by an adjective, and merge a preposition that is followed by a word. we utilizes PoS tagging approach to tag the transcription corpus. the tagged transcription is then used to find the new merged words.

A tag is a word property such as noun, pronoun, verb, adjective, adverb, preposition, conjunction, interjection, etc. Each language has its own tags. Tags may be different from language to language. In our method, we used the Arabic module of Stanford tagger (Stanford Log-linear Part-Of-Speech Tagger, 2011). The total number of tags of this tagger is 29 tags, only 13 tags were used in our method as listed in Table 3. As we mentioned, we focused on three kinds of tags: noun, adjectives, and preposition. In Table 3, DT is a shorthand for the determiner article (ال التعريف) that corresponds to "the" in English.

Table 3. A partial list of Stanford Tagger's tag with examples

#	Tag	Meaning	Example
1	ADJ_NUM	Adjective, Numeric	الرابعة السابع،
2	DTJJ	DT + Adjective	الجديد النفطية،
3	DTJJR	Adjective, comparative	العليا الكبرى،
4	DTNN	DT + Noun, singular or mass	المنظمة، العاصمة
5	DTNNP	DT + Proper noun, singular	القاهرة العراق،
6	DTNNS	DT + Noun, plural	السيارات، الولايات
7	IN	Preposition subordinating conjunction	حرف جر مثل : في حرف مصدري مثل :أنْ
8	JJ	Adjective	قيادية جديدة،
9	JJR	Adjective, comparative	كبرى أدنى،
10	NN	Noun, singular or mass	إنتاج، نجم
11	NNP	Proper noun, singular	لبنان أوبك،
12	NNS	Noun, plural	طلبات توقعات،
13	NOUN_QUANT	Noun, quantity	الربع، ثلثي

In this work, we used the Noun-Adjective as shorthand for a compound word generated by merging a noun and an adjective. We also used Preposition-Word as shorthand for a compound word generated by merging a preposition with a subsequent word. The prepositions used in our method include:

(من ، الى ، عن ، على ، في ، حتى ، منذ) (mundhu, Hata, fy, 'ala, 'an, 'ila, min), Other prepositions were not included as they are rarely used in MSA. Table 4 shows the tagger output for a simple non-diacritized sentence.

Thus, the tagger output is used to generate compound words by searching for Noun-Adjective and Preposition-Word sequences. Figure 9 shows two possible compound words: (بَرنَامِجضَخم) and (فِيالأُردُن) for Noun-Adjective case and for Preposition-Word case, respectively. These two compound words are, then, represented in new sentences as illustrated in Figure 9. Therefore, the three sentences (the original and the new ones) will be used, with all other cases, to produce the enhanced language model and the enhanced pronunciation dictionary.

Table 4. An Arabic sentence and its tags

An input sentence to the tagger	وأوضح عضو لجنة المقاولين في غرفة الرياض بشير العظم wa 'wdaHa 'udwu lajnata 'lmuqawilyna fy ghurfitu 'lriyaD bashyru 'l' aZm
Tagger output (read from left to right)	وأوضح/VBD عضو/NN لجنة/NN المقاولين/DTNNS في/IN غرفة/NN الرياض/DTNNP بشير/NNP العظم/DTNN

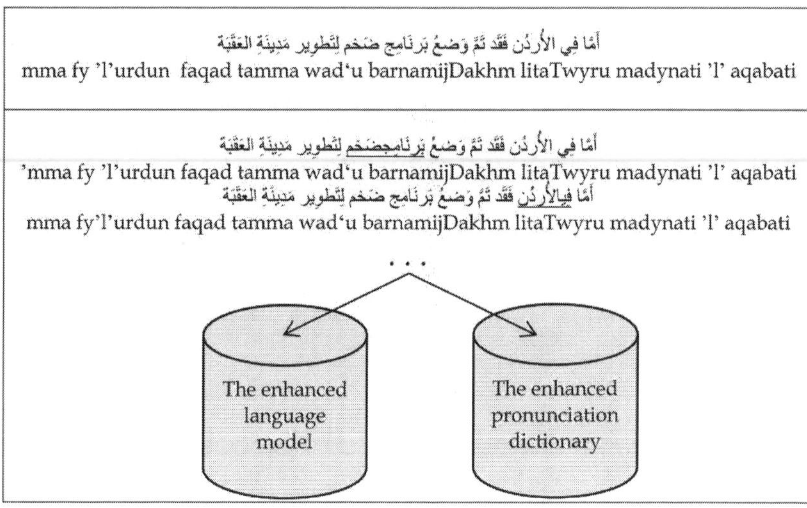

Figure 9. The compound words representations

Figure 10 shows the process of generating a compound word. It demonstrates that a noun followed by an adjective will be merged to produce a one compound word. similarly, the preposition followed by a word will be merged to perform a one compound word. It is noteworthy to mention that our method is independent from handling pronunciation variations that may occur at words junctures. That is, our method does not consider the phonological rules that could be implemented between certain words.

The steps for modeling cross-word phenomenon can be described by the algorithm (pseudocode) shown in Figure 11. In the figure, the Offline stage means that the stage is implemented once before decoding, while Online stage means that this stage needs to be repeatedly implemented after each decoding process.

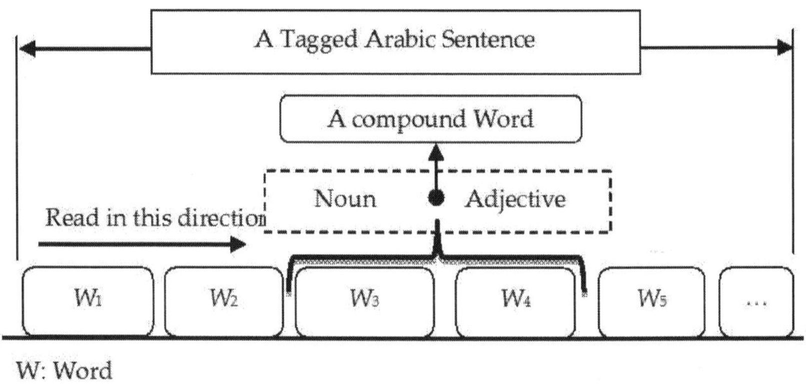

Figure 10. A Noun-Adjective compound word generation

```
                    Offline Stage
Using a PoS tagger, have the transcription corpus tagged
For all tagged sentences in the transcription file
      For each two adjacent tags of each tagged sentence
           If the adjacent tags are adjective/noun or word/preposition
                  Generate the compound word
                  Represent the compound word in the transcription
           End if
    End for
End for
Based on the new transcription, build the enhanced dictionary
Based on the new transcription, build the enhanced language model
                    Online Stage
Switching the variants back to its original separated words
```

Figure 11. Cross-word modeling algorithm using PoS tagging

7. THE RESULTS

The proposed method was investigated on a speaker-independent modern standard Arabic speech recognition system using Carnegie Mellon University Sphinx speech recognition engine. Three performance metrics were used to measure the performance enhancement: the word error rate (WER), out of vocabulary (OOV), and perplexity (PP).

WER is a common metric to measure performance of ASRs. WER is computed using the following formula:

$$WER = \frac{S + D + I}{N} \quad (2)$$

Where:

- S is the number of substituted words,
- D is the number of deleted words,
- I is the number of inserted words,
- N is the total number of words in the testing set.

The word accuracy can also be measured using WER as the following formula:

$$\text{Word Accuracy} = 1 - \text{WER}$$

OOV is a metric to measure the performance of ASRs. OOV is known as a source of recognition errors, which in turn could lead to additional errors in the words that follow (Gallwitz et al., 1996). Hence fore, increasing OOVs plays a significant role in increasing WER and deteriorating performance. In this research work, the baseline system is based on a closed vocabulary. The closed vocabulary assumes that all words of the testing set are already included in the dictionary. (Jurafsky & Martin, 2009) explored the differences between open and closed vocabulary. In our method, we calculate OOV as the percentage of recognized words that are not belonging to the testing set, but to the training set. The following formula is used to find OOV:

$$\text{OOV (baseline system)} = \frac{\text{none testing set words}}{\text{total words in the testing set}} * 100 \quad (3)$$

The perplexity of the language model is defined in terms of the inverse of the average log likelihood per word (Jelinek, 1999). It is an indication of the average number of words that can follow a given word, a measure of the predictive power of the language model, (Saon & Padmanabhan,2001). Measuring the perplexity is a common way to evaluate N-gram language model. It is a way to measure the quality of a model independent of any ASR system. Of course, The measurement is performed on the testing set. A lower perplexity system is considered better than one of higher perplexity. The perplexity formula is:

$$PP(W) = \sqrt[N]{\frac{1}{P(w_1, w_2, \ldots, w_N)}} \quad (4)$$

Where PP is the perplexity, P is the probability of the word set to be tested $W = w_1, w_2, \ldots, w_N$, and N is the total number of words in the testing set.

The performance detection method proposed by Plötz in (Plötz,2005) is used to investigate the achieved recognition results. A 95% is used as a level of confidence. The WER of the baseline system (12.21 %) and the total number of words in the testing set (9288 words) are used to find the confidence interval [εl, εh]. The boundaries of the confidence interval are found to be [12.21 – 0.68, 12.21 + 0.68] [11.53,12.89]. If the changed classification error rate is outside this interval, this change can be interpreted as statistically significant. Otherwise, It is most likely caused by chance.

Table 5 shows the enhancements for different experiments. Since the enhanced method (in Noun-Adjective case) achieved a WER of (9.82%) which is out of the above mentioned confidence interval [11.53,12.89], it is concluded that the achieved enhancement is statistically significant. The other cases are similar, i.e. (Preposition-Word, and Hybrid cases also achieved a significant improvement).

Table 5. Accuracy achieved and WERs for different cases

#	Experiment	Accuracy (%)	WER (%)	Enhancement (%)
	Baseline system	87.79	12.21	----------
1	Noun-Adjective	90.18	9.82	2.39
2	Preposition-Word	90.04	9.96	2.25
3	Hybrid (1 & 2)	90.07	9.93	2.28

Table 5 shows that the highest accuracy achieved is in Noun-Adjective case. The reduction in accuracy in the hybrid case is due to the ambiguity introduced in the language model. For more clarification, our method depends on adding new sentences to the transcription corpus that is used to build the language model. Therefore, adding many sentences will finally cause the language model to be biased to some n-grams (1-grams, 2-grams, and 3-grams) on the account of others.

The common way to evaluate the N-gram language model is using perplexity. The perplexity for the baseline is 34.08. For the proposed cases, the language models' perplexities are displayed in Table 6. The measurements were taken based on the testing set, which contains 9288 words. The enhanced cases are clearly better as their perplexities are lower. The reason for the low perplexities is the specific domains that we used in our corpus, i.e. economics and sports.

The OOV was also measured for the performed experiments. Our ASR system is based on a closed vocabulary, so we assume that there are no unknown words. The OOV was calculated as the percentage of recognized words that do not belong to the testing set, but to the training set. Hence,

$$\text{OOV (baseline system)} = \frac{\text{none testing set words}}{\text{total words in the testing set}} * 100 \qquad (5)$$

which is equal to 328/9288*100= 3.53%. For the enhanced cases, Table 6 shows the resulting OOVs. Clearly, the lower the OOV the better the performance is, which was achieved in all three cases.

Table 7 shows some statistical information collected during experiments. The "Total compound words" is the total number of Noun-Adjective cases found in the corpus transcription. The "unique compound words" indicates the total number of Noun-Adjective cases after removing duplicates. The last column, "compound words replaced" is the total number of compound words that were replaced back to their original two disjoint words after the decoding process and prior to the evaluation stage.

Despite the claim that the Stanford Arabic tagger accuracy is more than 96%, a comprehensive manual verification and correction were made on the tagger output. It was reasonable to review the collected compound words as our transcription corpus is small (39217 words). For large corpora, the accuracy of the tagger is crucial for the results. Table 8 shows an error that occurred in the tagger output. The word, for example, "وقال"(waqala) should be VBD instead of NN.

Table 6. Perplexities and OOV for different experiments

#	Experiment	Perplexity	OOV (%)
	Baseline System	34.08	328/9288 = 3.53%
1	Noun-Adjective	3.00	287/9288 = 3.09%
2	Preposition-Word	3.22	299/9288 = 3.21%
3	Hybrid (1 & 2)	2.92	316/9288 = 3.40%

Table 7. Statistical information for compound words

#	Experiment	Total compound words	unique compound words	compound words replaced
1	Noun-Adjective	3328	2672	377
2	Preposition-Word	3883	2297	409
3	Hybrid (1 & 2)	7211	4969	477

Table 8. Example of Stanford Arabic Tagger Errors

Sentence to be tagged	هذا وقال رئيس لجنة الطاقة بمجلس النواب ورئيس الرابطة الروسية للغاز إن الاحتكارات الأوروبية hadha waqala ra'ysu lajnati 'lTaqa bimajlisi 'lnuwab wa ra'ysu 'lrabiTa 'lrwsiya llghaz 'ina 'l'iHtikarati 'liwrobiya
Stanford Tagger output (read from left to right)	هذا/DT وقال/NN رئيس/NN لجنة/NN الطاقة/DTNN بمجلس/NN النواب/DTNN ورئيس/NN الرابطة/DTNN الروسية/DTJJ للغاز/NNP إن/NNP الاحتكارات/DTNNS الأوروبية/DTJJ

Table 9 shows an illustrative example of the enhancement that was achieved in the enhanced system. It shows that the baseline system missed one word "من"(min) while it appears in the enhanced system. Introducing a compound word in this sentence avoided the misrecognition that occurred in the baseline system.

According to the proposed algorithm, each sentence in the enhanced transcription corpus can have a maximum of one compound word, since sentences are added to the enhanced corpus once a compound word is formed. Finally, After the decoding process, the results are scanned in order to decompose the compound words back to their original form (two separate words). This process is performed using a lookup table such as:

الكُوَيتالدُّوَلِيّ الكُوَيت الدُّوَلِيّ ('lkuwaytldawly 'lkuwayt 'ldawly)

فِيمَطَارِ فِي مَطَارِ (fymatari fy matari)

8. DISCUSSION

Table 10 shows comparison results of the suggested methods for cross-word modeling. It shows that PoS tagging approach outperform the other methods (i.e. the phonological rules and small word merging) which were investigated on the same pronunciation corpus. The use of phonological rules was demonstrated in (AbuZeina et al. 2011a) while merging of small-words method was presented in (AbuZeina et al. 2011b). even though PoS tagging seems to be better than the other methods, more research should be carried out for more confidence. So, the comparison demonstrated in Table 10 is subject to change as more cases need to be investigated for both techniques. That is, cross-word was modeled using only two Arabic phonological rules, while only two compounding schemes were applied in PoS tagging approach.

The recognition time is compared with the baseline system. The comparison includes the testing set which includes 1144 speech files. The specifications of the machine where we conducted the experiments were as follows: a desktop computer which contains a single processing chip of 3.2GHz and 2.0 GB of RAM. We found that the recognition time for the enhanced method is almost the same as the recognition time of the baseline system. This means that the proposed method is almost equal to the baseline system in term of time complexity.

Table 9. An example of enhancement in the enhanced system

The text of a speech file to be tested	فِي المَرحَلَةِ السَّابِعَةِ وَالثَّلاثِين مِن الدَّورِيِّ الإسبَانِيِّ لِكُرَةِ القَدَم fy 'lmarHalati 'lsabi' a wa 'lthalathyn mina 'ldawry 'l'sbany likurati 'lqadam
As recognized by the baseline system	فِي المَرحَلَةِ السَّابِعَةِ وَالثَّلاثِين الدَّورِيِّ الإسبَانِيِّ لِكُرَةِ القَدَم fy 'lmarHalati 'lsabi' a wa 'lthalathyn mina 'ldawry 'l'sbany likurati 'lqadam
As recognized by the enhanced system	فِي المَرحَلَةِ السَّابِعَةِ وَالثَّلاثِين مِن الدَّورِيّالإسبَانِيِّ لِكُرَةِ القَدَم fy 'lmarHalati 'lsabi' a wa 'lthalathyn mina 'ldawry 'l'sbany likurati 'lqadam
Final output after decomposing the merging	فِي المَرحَلَةِ السَّابِعَةِ وَالثَّلاثِين مِن الدَّورِيِّ الإسبَانِيِّ لِكُرَةِ القَدَم fy 'lmarHalati 'lsabi' a wa 'lthalathyn mina 'ldawry 'l'sbany likurati 'lqadam

Table 10. Comparison between cross-word modeling techniques

#	System	Accuracy (%)	Execution Time (minutes)
	Baseline system	87.79	34.14
1	phonological rules	90.09	33.49
2	PoS tagging	90.18	33.05
3	small word merging	89.95	34.31
4	Combined system (1,2,and3)	88.48	30.31

9. FURTHER RESEARCH

As future work, we propose investigating more word-combination cases. In particular, we expect that the construct phrases Idafa (الإضافة) make a good candidate. Examples include: (سلسلة جبال, silsilt jibal), (مطار بيروت, maTaru bayrwt), (مدينة القدس, madynatu ’lquds). Another suggested candidate is the Arabic "and" connective (واو العطف), such as: (مواد أدبية ولغوية, mawad ’dabiyah wa lughawiyah), (yata‘allaqu biqaDaya ’l ‘iraqi wa ‘lsudan يتعلق بقضايا العراق والسودان،). A hybrid system could also be investigated. It is possible to use the different cross-word modeling approaches in a one ASR system. It is also worthy to investigate how to model the compound words in the language model. In our method, we create a new sentence for each compound word. we suggest to investigate representing the compound word exclusively with its neighbors. for example, instead of having two complete sentences to represent the compound words (بَرنَامِجضَخم, barnamijDakhm) and (فِيالأُردُن, fy’l’urdun) as what we proposed in our method:

أَمَّا فِي الأُردُن فَقَد تَمَّ وَضعُ بَرنَامِجضَخم لِتَطوِير مَدِينَةِ العَقَبَة

’mma fy ’l’urdun faqad tamma wad‘u barnamijDakhm litaTwyru madynati ’l ’aqabati

أَمَّا فِيالأُردُن فَقَد تَمَّ وَضعُ بَرنَامِج ضَخم لِتَطوِير مَدِينَةِ العَقَبَة

’ mma fy’l’urdun faqad tamma wad‘u barnamijDakhm litaTwyru madynati ’l ’aqabati

We propose to add the compound words only with their adjacent words like:

وَضعُ بَرنَامِجضَخم لِتَطوِير

waD‘u barnamijDakhm litaTwyr

أَمَّا فِيالأُردُن فَقَد

’ mma fy ’l’urdun faqad

A comprehensive research work should be made to find how to effectively represent the compound words in the language model. In addition, we highly recommend further research in PoS tagging for Arabic.

10. CONCLUSION

The proposed knowledge-based approach to model cross-word pronunciation variations problem achieved a feasible improvement. Mainly, PoS tagging approach was used to form compound words. The experimental results clearly showed that forming compound words using a noun and an adjective achieved a better accuracy than merging of a preposition and its next word. The significant enhancement we achieved has not only come from the cross-word pronunciation modeling in the dictionary, but also indirectly from the recalculated n-grams probabilities in the language model. We also conclude that Viterbi algorithm works better with long words. Speech recognition research should consider this fact when designing dictionaries. We found that merging words based on their types (tags) leads to significant improvement in Arabic ASRs. We also found that the proposed method outperforms the other cross-word methods such as phonological rules and small-words merging.

ACKNOWLEDGEMENT

The authors would like to thank King Fahd University of Petroleum and Minerals for providing the facilities to write this chapter. We also thank King Abdulaziz City for Science and Technology (KACST) for partially supporting this research work under Saudi Arabia Government research grant NSTP # (08-INF100-4).

REFERENCES

1. M. A. Abushariah, A. M. , R. N. Ainon, R. Zainuddin, M. Elshafei, O. O. Khalifa, O.Arabic speaker-independent continuous automatic speech recognition based on a phonetically rich and balanced speech corpus. Int. Arab J. Inf. Technol., 2012201298493

2. AbuZeina D., Al-Khatib W., Elshafei M., "Small-Word Pronunciation Modeling for Arabic Speech Recognition: A Data-Driven Approach", Seventh Asian Information Retrieval Societies Conference, Dubai, 2011b.

3. AbuZeina D., Al-Khatib W., Elshafei M., Al-Muhtaseb H., "Cross-word Arabic pronunciation variation modeling for speech recognition", International Journal of Speech Technology,2011a.

4. M. Afify, L. Nguyen, B. Xiang, S. Abdou, J. Makhoul, Recent progress in Arabic broadcast news transcription at BBN. In: Proceedings of INTERSPEECH. 200516371640

5. M. Alghamdi, M. Elshafei, H. Almuhtasib, 2009Arabic broadcast news transcription system. Int J Speech Tech 10183195

6. M. Ali, M. Elshafei, M. Alghamdi, H. Almuhtaseb, A. Alnajjar, . Arabic, Dictionaries. Phonetic, Speech. for, Recognition, Journal of Information Technology Research, 2420096780

7. Alotaibi YA2004Spoken Arabic digits recognizer using recurrent neural networks. In: Proceedings of the fourth IEEE international symposium on signal processing and information technology, 195199

8. F. Al-Otaibi, 2001speaker-dependant continuous Arabic speech recognition. M.Sc. thesis, King Saud University

9. I. Amdal, E. Fosler-Lussier, 2003Pronunciation variation modeling in automatic speech recognition. Telektronikk, 2.2003, 7082

10. M. Azmi, H. Tolba, S. Mahdy, Fashal, M(2008Syllable-based automatic Arabic speech recognition in noisy-telephone channel. In: WSEAS transactions on signal processing proceedings, World Scientific and Engineering Academy and Society (WSEAS), 44211220

11. H. Bahi, M. Sellami, 2001Combination of vector quantization and hidden Markov models for Arabic speech recognition. ACS/IEEE international conference on computer systems and applications, 2001

12. M. Benzeghiba, R. De Mori, et al.2007Automatic speech recognition and speech variability: a review. Speech Commun 49(10-11):763 EOF786 EOF

13. J. Billa, M. Noamany, et al.2002Audio indexing of Arabic broadcast news. 2002 IEEE international conference on acoustics, speech, and signal processing (ICASSP)

14. H. Bourouba, R. Djemili, et al.2006New hybrid system (supervised classifier/HMM) for isolated Arabic speech recognition. 2nd Information and Communication Technologies, 2006. ICTTA'06

15. F. Choi, S. Tsakalidis, et al.2008Recent improvements in BBN's English/Iraqi speech-to-speech translation system. IEEE Spoken language technology workshop, 2008. SLT 2008

16. P. Clarkson, R. Rosenfeld, 1997Statistical language modeling using the CMU-Cambridge toolkit. In: Proceedings of the 5th European conference on speech communication and technology, Rhodes, Greece.

17. M. Elmahdy, R. Gruhn, et al.2009Modern standard Arabic based multilingual approach for dialectal Arabic speech recognition. In: Eighth international symposium on natural language processing, 2009. SNLP'09

18. Elmisery FA, Khalil AH et al2003A FPGA-based HMM for a discrete Arabic speech recognition system. In: Proceedings of the 15th international conference on microelectronics, 2003. ICM 2003

19. A. Emami, L. Mangu, 2007Empirical study of neural network language models for Arabic speech recognition. IEEE workshop on automatic speech recognition and understanding, 2007. ASRU

20. Essa EM, Tolba AS et al2008A comparison of combined classifier architectures for Arabic speech recognition. International conference on computer engineering and systems, 2008. ICCES 2008

21. A. Farghaly, K. Shaalan, 2009Arabic natural language processing: challenges and solutions. ACM Trans Asian Lang Inform Process 84122

22. M. J. F. Gales, F. Diehl, et al.2007Development of a phonetic system for large vocabulary Arabic speech recognition. IEEE workshop on automatic speech recognition and understanding, 2007. ASRU

23. F. Gallwitz, E. Noth, et al.1996A category based approach for recognition of out-of-vocabulary words. In: Proceedings of fourth international conference on spoken language, 1996. ICSLP 96

24. Hwang M-H1993Subphonetic acoustic modeling for speaker-independent continuous speech recognition, Ph.D. thesis, School of Computer Science, Carnegie Mellon University.

25. H. Hyassat, Zitar. R. Abu, 2008Arabic speech recognition using Sphinx engine. Int J Speech Tech 9(3-4):133 EOF150 EOF

26. T. Imai, A. Ando, et al.1995A new method for automatic generation of speaker-dependent phonological rules. 1995 international conference on acoustics, speech, and signal processing, 1995. ICASSP-95

27. F. Jelinek, 1999Statistical methods for speech recognition, Language, speech and communication series. MIT, Cambridge, MA

28. D. Jurafsky, J. Martin, 2009Speech and language processing, 2nd edn. Pearson, NJ

29. M. Khasawneh, K. Assaleh, et al.2004The application of polynomial discriminant function classifiers to isolated Arabic speech recognition. In: Proceedings of the IEEE international joint conference on neural networks, 2004

30. K. Kirchhofl, J. Bilmes, S. Das, N. Duta, M. Egan, G. Ji, F. He, J. Henderson, D. Liu, M. Noamany, P. Schoner, R. Schwartz, D. Vergyri, 2003Novel approaches to Arabic speech recognition: report from the 2002 John-Hopkins summer workshop, ICASSP 2003, I344I347

31. H. J. Kuo, L. Mangu, et al.2010Morphological and syntactic features for Arabic speech recognition. 2010 IEEE international conference on acoustics speech and signal processing (ICASSP)

32. L. Lamel, A. Messaoudi, et al.2009Automatic speech-to-text transcription in Arabic. ACM Trans Asian Lang Inform Process 8411822Arabic Speech Recognition Systems

33. Lee KF1988Large vocabulary speaker independent continuous speech recognition: the Sphinx system. Doctoral dissertation, Carnegie Mellon University.

34. A. Messaoudi, J. L. Gauvain, et al.2006Arabic broadcast news transcription using a one million word vocalized vocabulary. 2006 IEEE international conference on acoustics, speech and signal processing, 2006. ICASSP 2006 proceedings

35. Mokhtar MA, El-Abddin AZ1996A model for the acoustic phonetic structure of Arabic language using a single ergodic hidden Markov

model. In: Proceedings of the fourth international conference on spoken language, 1996. ICSLP 96

36. G. Muhammad, Malki. K. Al, et al.2011Automatic Arabic digit speech recognition and formant analysis for voicing disordered people. 2011 IEEE symposium on computers and informatics (ISCI)

37. M. Nofal, Reheem. E. Abdel, et al.2004The development of acoustic models for command and control Arabic speech recognition system. 2004 international conference on electrical, electronic and computer engineering, 2004. ICEEC'04

38. Rambow. Owen, Chiang. David, et al.Arabic. Parsing, Final. Dialects, . Report-Version, January, 2006

39. http://old-site.clsp.jhu.edu/ws05/groups/arabic/documents/final-report.pdf

40. J. Park, F. Diehl, et al.2009Training and adapting MLP features for Arabic speech recognition.IEEE international conference on acoustics, speech and signal processing, 2009. ICASSP 2009

41. T. Plötz, 2005Advanced stochastic protein sequence analysis, Ph.D. thesis, Bielefeld University

42. Rabiner, L. R. and Juang, B. H., Statistical Methods for the Recognition and Understanding of Speech, Encyclopedia of Language and Linguistics, 2004.

43. Ryding KC2005A reference grammar of modern standard Arabic (reference grammars). Cambridge University Press, Cambridge.

44. A. Sagheer, N. Tsuruta, et al.2005Hyper column model vs. fast DCT for feature extraction in visual Arabic speech recognition. In: Proceedings of the fifth IEEE international symposium on signal processing and information technology, 2005

45. G. Saon, M. Padmanabhan, 2001Data-driven approach to designing compound words for continuous speech recognition. IEEE Trans Speech Audio Process 94327332

46. G. Saon, H. Soltau, et al.2010The IBM 2008 GALE Arabic speech transcription system. 2010 IEEE international conference on acoustics speech and signal processing (ICASSP)

47. H. Satori, M. Harti, N. Chenfour, 2007Introduction to Arabic speech recognition using CMU Sphinx system. Information and communication technologies international symposium proceeding ICTIS07, 2007

48. Selouani S-A, Alotaibi YA2011Adaptation of foreign accented speakers in native Arabic ASR systems. Appl Comput Informat 91110

49. M. Shoaib, F. Rasheed, J. Akhtar, M. Awais, S. Masud, S. Shamail, 2003A novel approach to increase the robustness of speaker independent Arabic speech recognition. 7th international multi topic conference, 2003. INMIC 2003. 8-9 Dec 2003, 371376

50. R. Singh, B. Raj, et al.2002Automatic generation of subword units for speech recognition systems." Speech and Audio Processing, IEEE Transactions on 1028999

51. H. Soltau, G. Saon, et al.2007The IBM 2006 Gale Arabic ASR system. IEEE international conference on acoustics, speech and signal processing, 2007. ICASSP 2007

52. Stanford Log-linear Part-Of-Speech Tagger,2011http://nlp.stanford.edu/software/tagger.shtml

53. M. Taha, T. Helmy, et al.2007Multi-agent based Arabic speech recognition. 2007 IEEE/WIC/ ACM international conferences on web intelligence and intelligent agent technology workshops

54. The CMU Pronunciation Dictionary2011http://www.speech. cs.cmu.edu/cgi-bin/cmudict,Accessed 1 September 2011.

55. D. Vergyri, K. Kirchhoff, K. Duh, A. Stolcke, 2004Morphology-based language modeling for Arabic speech recognition. International conference on speech and language processing. Jeju Island, 12521255

56. B. Xiang, K. Nguyen, L. Nguyen, R. Schwartz, J. Makhoul, 2006Morphological ecomposition for Arabic broadcast news transcription. In: Proceedings of ICASSP, vol I. Toulouse, 10891092

Chapter 8

Optimizing Automatic Speech Recognition for Low-Proficient Non-Native Speakers

Joost van Doremalen, Catia Cucchiarini, and Helmer Strik

Department of Language and Speech, Radboud University,
6500 HD Nijmegen, The Netherlands

1. INTRODUCTION

The increasing demand for innovative applications that support language learning has led to a growing interest in Computer-Assisted Language Learning (CALL) systems that make use of ASR technology. Such systems can address oral proficiency, one of the most problematic skills in terms of time investments and costs, and are seriously being considered as a viable alternative to teacher-fronted lessons. However, developing ASR-based CALL systems that can provide training and feedback for second language (L2) speaking is not trivial.

First of all, because non-native speech is atypical in many respects and, as such, it poses serious problems to ASR systems [1, 2, 3, 4]. Non-native speech may deviate from native speech with respect to pronunciation, morphology, syntax, and the lexicon. Pronunciation is considered a difficult skill to learn in a second language (L2), and even highly proficient non-native speakers often maintain a foreign accent [5]. An important limiting factor in acquiring the pronunciation of an L2 is considered to be interference from the first language (L1). As a consequence, the pronunciation of non-native speakers may deviate in various respects and to different degrees from that of native speakers. Deviations may concern prosodic or segmental aspects of speech or both. At the segmental level, the deviations may be limited to phonetic properties without really compromising phonemic distinctions, or they may blur phonemic distinctions and thus have more serious consequences for intelligibility. For instance, non-native speakers may use phonemes from their L1 when speaking the target language [5] or they may have difficulties in perceiving and/or realizing phonetic contrasts that are not distinctive in their mother tongue. Illustrations of this phenomenon are provided by Italian speakers of English who realize English /p/, /t/, /k/, /b/, /d/, and /g/ with voice onset time (VOT) values that differ from those employed by native speakers [5]. Such deviations might cause misunderstandings in certain cases, but do not necessarily hamper communication because the distinction between separate phonemes, that is, /p/ versus /b/ in the target language is preserved, albeit differently realized. Native speakers will probably perceive the difference and consider it as foreign accent. More problematic deviations may arise when the difficulty in perceiving and realizing phonetic features of the target language that are not distinctive in the mother tongue leads non-native speakers to blur the distinction between phonemes in the target language, thus producing one phoneme instead of two distinct ones. This is the case with many non-native speakers of English, for instance, Germans [6], who have difficulty in realizing the distinction between the English phonemes /ae/ and /e/ and often produce /e/ when /ae/ should be used, or Japanese speakers of English who have difficulty in distinguishing /l/ and /r/ [7] and may end up producing sounds that are neither an English /l/ nor an English /r/. In such cases, confusion may arise because distinct words will be realized in the same way. This can also happen when speech sounds are inappropriately deleted or inserted, which is another common phenomenon in non-native speech [8].

With respect to morphology and syntax the speech of non-natives may also exhibit deviations from that of native speakers [9]. At the level of morphology, they may find it difficult to produce correct forms of verbs, nouns, adjectives, articles, and so forth, especially when the morphological distinction hinges on subtle phonetic distinctions, such as the presence of a plosive or fricative sound in consonant clusters or the distinction between two similar vowels (lead versus led). Irregular verbs and nouns may also pose serious problems, resulting in the production of nonexistent regularized forms. Deviations in syntax may concern the structure of sentences, the ordering of constituents and their omission or insertion. As to vocabulary, non-native speakers also tend to have a limited and often deviant lexicon. Finally, non-native speech exhibits more disfluencies and hesitation phenomena than native speech and is characterized by a lower speech rate [10, 11, 12, 13, 14].

All these problems are compounded when dealing with speech of non-natives that are still in the process of learning the language. In general, the degree of deviation from native speech and the incidence of disfluencies will be in inverse relation to the degree of proficiency in the target language. Considering that ASR-based CALL systems are intended for L2 learners, including beginner and intermediate learners, it follows that the type of non-native speech that has to be handled in this context is, in general, even more atypical and, therefore, more challenging, than the non-native speech that is usually encountered in other ASR applications that do not have such a teaching function, like information systems or access interfaces.

To circumvent the ASR problems caused by non-native speech, various techniques have been proposed to restrict the search space and make the task easier. A major distinction can be drawn between strategies that are essentially aimed at constraining the output of the learner so that the speech becomes more predictable and techniques that are aimed at improving the decoding of non-native speech. Such strategies are often used in combination.

Within the first category, a possible strategy consists in eliciting constrained output from learners by letting them read aloud an utterance from a limited set of answers presented on the screen or by allowing a limited amount of freedom in formulating responses, as in the *Subarashii* [15] and the *Let's Go* systems [16]. More freedom in user responses is particularly necessary in ASR-based CALL systems that are intended for practicing grammar in speaking proficiency. While for practicing pronunciation it may suffice to read sentences aloud, to practice grammar learners need to have some freedom in formulating answers in order to show whether they are able to produce correct forms. Less constrained output is not only problematic because it is more difficult to predict but also because, in general, it is accompanied by a higher incidence of disfluencies and hesitations. In a study on read and spontaneous speech produced by non-native speakers of Dutch [12], we found that extemporaneous speech contains many more filled pauses and disfluencies than read speech. The more freedom is allowed to the learner, the more complex the recognition task will be. In addition, tasks with more freedom will in general be characterized by a higher cognitive load, which, in turn, is likely to lead to more disfluencies being produced [17], thus making the recognition task even more difficult.

The second category of techniques for dealing with non-native speech, that is, those that are aimed at improving decoding, comprises methods for optimizing the acoustic models, the lexicon, and the language model in order to compensate for the deviations in pronunciation, morphology, and syntax.

All the factors mentioned above make it clear that to develop ASR-based CALL systems for oral proficiency it is necessary to take measures at different levels. A first important measure consists in designing exercises that allow some freedom to the learners in producing answers, but that are predictable enough to be handled by ASR. How much freedom can be allowed is of course dependent on the quality of decoding.

These are exactly the problems we face in the DISCO project, which is aimed at developing a prototype of an ASR-based CALL application for practicing oral skills in Dutch as a second language (DL2) and providing intelligent feedback on important aspects of speaking performance such as pronunciation, morphology, and syntax. The application should be able to detect and give feedback on errors that are made by learners of DL2 at the A2 level of the Common European Framework (CEF). This is achieved by generating a predefined list of possible (correct and incorrect) responses for each exercise.

In this project we intend to use a two-step procedure in which first the content of the utterance is determined (what was said), and subsequently the form of the utterance is analysed (how it was said). In the first (recognition) step the system should tolerate deviations in the way utterances are spoken, while in the second (error detection) step, strictness is required (see also [18, 19]). In the first step of the two-step procedure, two phases can be distinguished, (a) utterance selection, and (b) utterance verification (UV). When learners are allowed some freedom in formulating their responses, there is always the possibility that the learner's response is not present in the predefined list and is recognized incorrectly in phase (a) as one of the utterances of the predefined list. Also, even if the utterance is present in the list, it can also be recognized incorrectly. Giving feedback on the basis of an incorrectly recognized utterance is confusing and thus should be avoided. Therefore, utterance verification (UV) is carried out in phase (b).

In this paper we present two experiments we carried out in order to test both utterance selection and utterance verification for our system using state-of-the-art techniques. In the utterance selection phase one of the utterances from the predefined list is selected, and in the utterance verification phase it is determined whether this utterance should be passed on to the following stages of the CALL system (error detection, feedback, etc.). While in the final system both phases should work in tandem, we studied (optimized, evaluated, etc.) the two phases in isolation, for diagnostic purposes, to acquire a better understanding, and thus, finally, to obtain a better functioning system.

In Section 2 we discuss related work on non-native speech recognition and utterance verification. In Section 3, we introduce our system architecture and relate the choices for the experimental settings to previous work. In Sections 4 and 5, we present two experiments that are aimed at optimizing and evaluating utterance selection and utterance verification using realistic test material. In Section 6, we discuss the results of the two experiments in combination and consider the implications for our CALL application.

2. RELATED WORK

In automatic speech recognition (ASR) the recognition result is often obtained using the *maximum a posteriori*(MAP) decision rule decoder:

$$\hat{w} = \arg\max_{w \in W} p(w \mid \mathbf{x}), \tag{1}$$

where $p(w \mid \mathbf{x})$ is the posterior probability of a word sequence w in a set of word sequences W given a sequence of acoustic observations $\mathbf{x}$ and $\hat{w}$ is the recognition result that maximizes the posterior probability.

By using Bayes rule (1) can be reformulated as (2), and given that $\mathbf{x}$ is the same for all word sequences in W , it can be rewritten as (3):

$$\hat{w} = \arg\max_{w \in W} \frac{p(\mathbf{x} \mid w)p(w)}{p(\mathbf{x})} \tag{2}$$

$$= \arg\max_{w \in W} p(\mathbf{x} \mid w)p(w). \tag{3}$$

By implementing (3), we can still find the optimal sequence of words $\hat{W}$ in W. However, it is generally not only important to find the best sequence of words $\hat{W}$ relative to the other sequences (see (3)) but also quantitatively assess the confidence in the recognition result in an absolute sense. This number is called the confidence measure (CM) of the recognition result and the problem of accepting or rejecting a recognition result is called utterance verification (UV).

Both (non-native) speech decoding and utterance verification are the key aspects of this research. We will now relate our research on both problems to other recent work.

2.1. Non-Native Speech Decoding

In the ASR community, it has long been known that the differences between native and non-native speech are so pervasive as to degrade ASR performance considerably (e.g., [1, 20, 21]). These differences affect essentially all three components of an ASR system. As explained in Section 1, non-natives often use different words and word orders (language model), produce sounds differently (acoustic models), pronounce words differently (lexicon) (see, e.g., [2]), and generally have a lower speech rate and produce more disfluencies [10, 11, 12]. A short overview of research on the three components of the ASR is provided in this section.

In attempts aimed at improving ASR performance on non-native speech, the acoustic models have received most attention. Various kinds of acoustic models can and have been used. First of all, it is possible to train acoustic models on speech material of the target language (L2). However, the recognition performance obtained with such models is usually not sufficient or at any rate considerably lower than the performance on native speech, because of the various deviations in the speech of non-natives [20, 21]. Models can also be obtained by training exclusively on non-native (L2) speech [22, 23], or on combinations of L1 and L2 speech. Regarding the latter, two different approaches can be adopted: "model merging" and "parallel models." In the "parallel models" approach, acoustic models for both languages are stored, and during decoding the recognizer determines which models fit the data better [24, 25, 26, 27]. In the "model merging" (or model interpolation) approach, acoustic models of both languages are combined, in order to obtain a new set of acoustic models [26]. The obvious disadvantage of these L1-L2 approaches is that they can only be applied to fixed L1-L2 pairs. An alternative approach that can be applied consists in employing adaptation techniques, such as the common Maximum Likelihood Linear Regression (MLLR) and MAP techniques, which have been shown to improve recognition performance [20, 21, 23, 26, 28].

Improving ASR performance on non-native speech can also be carried out at the level of the lexicon. An obvious way to model pronunciation variation at the level of the lexicon is by adding pronunciation variants to the lexicon [29, 30]. In the case of non-native speech these variants should reflect possible L1-induced mispronunciations of words L2 learners may produce [18, 31, 32]. These variants can be generated by means of rules obtained from studying non-native speech [18, 32]. Another possibility to generate non-native variants for an L2 lexicon is to apply an L1 phoneme recognizer to L2 speech [31]. The advantage of the latter approach is that no learner data are needed, but a disadvantage is that phoneme recognizers for all source languages (L1s) are needed. The work in [31] also carried out speaker adaptation, and the improvements they obtained with speaker adaptation were much larger than those obtained with lexicon adaptation.

The choices regarding the language model depend to a large extent on the design of the CALL system, the type of items present in the CALL system. In spoken CALL systems, use could be made of closed or open items. For instance, the learner could be asked to repeat an utterance that is spoken by the system, or read an utterance presented on the screen. In these cases, the required responses are known, which in turn makes it possible to derive specific language models for every item. Alternatively, in some cases, a language model might not be used at all, depending on the approach that is chosen. For more open items in a CALL system (e.g., a question, or a turn in the dialogue), a possibility is to try to elicit constrained responses. This makes it possible to activate a specific language model for every item containing only those utterances that are expected in that given context. In these cases, a "stricter" language model can be used [33, 34, 35]. In this way, recognition performance can again be maximized without affecting the face validity of the application. This is done, for instance, in the Auralog programs [36]. In spite of the constraints that are introduced to improve ASR performance, the students can still have the feeling that they are interacting with the system and that they have control over the conversation [36].

2.2. Utterance Verification

In the literature roughly three approaches for tackling the UV problem can be distinguished: (1) posterior probability estimation, (2) statistical hypothesis testing, and (3) confidence predictors. We will now give a short overview of these approaches (see [37] for a more detailed overview).

(1) One approach to CM is to directly estimate the posterior probability of the recognition result $\hat{W}$ given the acoustic observations $\mathbf{x}$:

$$p(\hat{W} \mid \mathbf{x}) = \frac{p(\mathbf{x} \mid \bar{W})p(\bar{W})}{p(\mathbf{x})}, \tag{4}$$

and reject the recognition result $\hat{W}$ when it is below a given threshold θ. The greatest challenge with respect to this approach is accurately estimating the denominator $p(\mathbf{x})$. One solution is to estimate it from a word lattice [38], and this generally provides a good result when the lattice contains enough word hypotheses. The lattice-based approach can be viewed as approximating the posterior probability where $p(\mathbf{x})$ is written as $\Sigma_i p(\mathbf{x} \mid W_i)p(W_i)$ and i ranges over all sequences of words in a pruned search space.

Another approach to estimating $\Sigma_i p(\mathbf{x} \mid W_i)p(W_i)$ is using a free phone recognizer (FPR) [39, 40] and approximate:

$$p(\mathbf{x}) \approx p(\mathbf{x} \mid u_{\mathrm{FPR}})p(u_{\mathrm{FPR}}), \tag{5}$$

where u_{FPR} is the optimal phone string found using a free phone recognizer.

(2) Another popular method to UV is statistical hypothesis testing, in which the null hypothesis H_0 states that the recognition result is a correct representation of the speech signal and the alternative hypothesis H_a states that the recognition result is not a correct representation. Then the criterion of accepting the null hypothesis becomes:

$$\frac{p(\mathbf{x} \mid \bar{W})}{p(\mathbf{x} \mid \neg\bar{W})} > \theta \tag{6}$$

in which the numerator equals the acoustic likelihood of $\hat{W}$, the denominator equals the acoustic likelihood of all sequences of words other than $\hat{W}$ (usually called the *antimodel*), and θ a predefined threshold. The main difficulty with this approach is defining and training the antimodel.

(3) Apart from estimating the posterior probability or statistical hypothesis testing, another method to UV is using predictors such as:

a) acoustic stability,

b) hypothesis density,

c) duration information,

and combining these using a machine learning model. Some machine learning techniques that have been used in the past are artifical neural networks (ANN), linear discriminant analysis (LDA) classifiers, and binary decision trees.

Acoustic stability [38] refers to stability of the recognition result given different weightings of the acoustic model and language model scores. When the recognition result remains stable given fluctuations in these weightings, it means that we can be more confident that it is correctly recognized. Hypothesis density [41] refers to the average density of the word lattice generated during decoding. When there are a lot of competing hypotheses in a pruned search space at each point in time this means that we can be less confident that the recognition result is correct. Duration modelling for UV usually comes down to capturing the amount of deviation of the phoneme durations in the recognition result from *normal* phone durations [42]. Deviating durations in the recognition result decreases the confidence that it is recognized correctly.

3. EXPERIMENTAL SYSTEM

In Figure 1, the architecture of our CALL system is shown. The input of the system is the learner's speech and a list of predicted responses in the form of transcriptions of sequences of words. Utterance selection is then performed to choose the best fitting (1-Best) response from this list. In the next phase the 1-Best response is verified. If the response is accepted, error detection on this response is carried out. Errors are detected on multiple levels, that is, syntax, morphology, and pronunciation. If the response is not accepted, the user is prompted to try again.

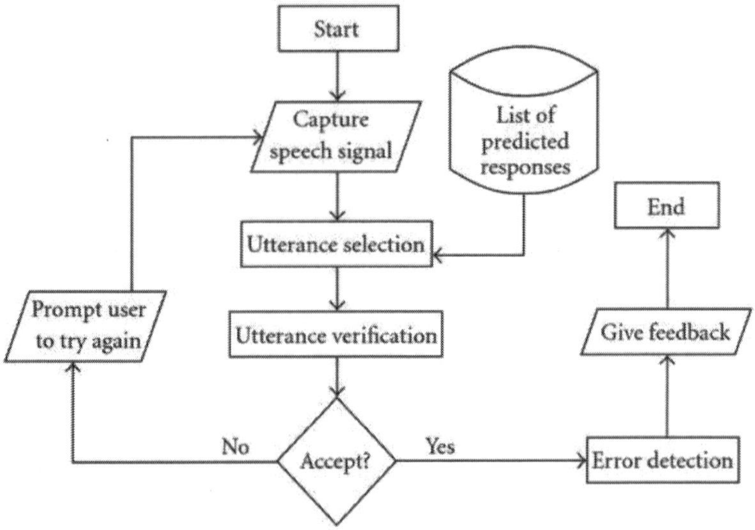

Figure 1. System architecture.

It is difficult for general Hidden Markov modelling methods to discriminate between utterances that are acoustically very similar [43]. Therefore, in the final CALL system we will probably use the following procedure: the output of the first step is a cluster of similar responses (e.g., according to a phonetically-based distance measure), and a more detailed analysis is carried out in the second (error detection) step to determine what was actually uttered and where to give feedback on.

We will now explain the main choices we made for our system regarding utterance selection and utterance verification procedures.

3.1. Utterance Selection

In the literature many approaches have already been proposed to improve the performance of speech recognition for non-natives. A large deal of the research concerned one or a small number of fixed (L1-L2) language pairs. In these approaches material of the source language (L1) or material for specific L1-L2 pairs was employed to enhance ASR for these language pairs.

However, since our system is intended for learners of Dutch with different mother tongues, approaches that require material of L1 or specific L1-L2 pairs are not feasible in our case for either of the three components of an ASR system (acoustic models, lexicon, and language model). Consequently, we made the following choices.

For the acoustic models we decided to start with training the acoustic models on Dutch native speech. Next, we used read speech of language learners (DL2 speech) to retrain the acoustic models (see Section 4.1.4). Such retraining of the acoustic models is also possible in a realistic CALL application, albeit not online, after a so-called enrolment phase, as used in dictation systems. Especially if the system has to be used extensively by a learner, it is possible to make it as suitable as possible for that specific learner. At the level of the lexicon we could not make use of L1 phoneme recognizers, as was done by [31], and thus we added pronunciation variants to the lexicon that were generated by means of data-derived rules (for further details, see Section 4.1.5). Finally, we decided to use specific language models for every item in the CALL system that are based on a list of predicted (correct and incorrect) responses (see Section 4.1.3).

3.2. Utterance Verification

In Section 2.2, we have given a short overview of the three key approaches to UV, that is, (1) posterior probability estimation, (2) statistical hypothesis testing, and (3) predictor combination. Most of these approaches are aimed at UV in large vocabulary tasks, that is, posterior probability estimation using word lattices and predictor features like acoustic stability and hypothesis density. Furthermore, training explicit antimodels for statistical hypothesis testing is conceptually and practically difficult for speakers with a large variety of L1 backgrounds [44]. For these reasons, we have chosen a form of predictor combination in which a likelihood ratio similar to (6) in statistical hypothesis testing is combined with phone durations. The rationale behind this choice is explained in detail in Section 5.1.2.

4. EXPERIMENT 1: UTTERANCE SELECTION

To goal of this experiment is to develop a procedure for selecting utterances from a list of predicted responses and to evaluate the effects of different language models, pronunciation lexicons, and acoustic models.

4.1. Method

4.1.1. Material

The speech material for the present experiments was taken from the JASMIN speech corpus [45], which contains speech of children, non-natives, and elderly people. Since the non-native component of the JASMIN corpus was collected for the aim of facilitating the development of ASR-based language learning applications, it is particularly suited for our purpose. Speech from speakers with different mother tongues was collected, because this realistically reflects the situation in Dutch L2 classes. These speakers have relatively low proficiency levels, namely, A1, A2, and B1 of the Common European Framework (CEF), because it is for these levels that ASR-based CALL applications appear to be most needed.

The JASMIN corpus contains speech collected in two different modalities: read speech and human-machine dialogues. The latter were used for our experiments because they more closely resemble the situation we will encounter in our CALL application. The JASMIN dialogues were collected through a Wizard-of-Oz-based platform and were designed such that the wizard was in control of the dialogue and could intervene when necessary. In addition, recognition errors were simulated and difficult questions were asked to elicit some typical phenomena of human-machine interaction that are known to be problematic in the development of spoken dialogue systems, such as hyperarticulation, restarts, filled pauses, self-talk, and repetitions.

The material we used for the present experiments consists of speech from 45 speakers, 40% male and 60% female, with 25 different L1 backgrounds. Ages range from 19 to 55, with a mean of 33. The speakers each give answers to 39 questions about a journey. We first deleted the utterances that contain crosstalk, background noise, and whispering from the corpus. After deletion of these utterances the material consists of 1325 utterances.

The mean signal-to-noise-ratio (SNR) of the material is 24.9 with a standard deviation of 5.1.

Considering all these characteristics, we can state that the JASMIN non-native dialogues are similar to the speech we will encounter in our CALL application for various reasons: (1) they contain answers to relatively constrained questions, (2) they contain semispontaneous speech, (3) of non-natives with different L1s, (4) which features spontaneous phenomena such as filled pauses and disfluencies. However, since hesitation phenomena were purposefully induced in the JASMIN dialogues, their incidence is probably higher than in typical non-native dialogues.

4.1.2. Speech Recognizer

The speech recognizer we used in this research is SPRAAK [46], an open source hidden markov model (HMM)-based ASR package. The input speech, sampled at 16 kHz, is divided into overlapping 32 milliseconds Hamming windows with a 10 milliseconds shift and preemphasis factor of 0.95. However, 12 Mel-frequency cepstral coefficients (MFCC) plus c_0, and their first and second order derivatives were calculated and cepstral mean subtraction (CMS) was applied. The constrained language models and pronunciation lexicons are implemented as finite state machines (FSM).

To simulate the ASR task in our CALL application, we generated lists of the answers given by each speaker to each of the 39 questions. These lists mimic the predicted responses in our CALL application task because they contain (a) responses to relatively closed questions and (b) morphologically and syntactically correct and incorrect responses.

4.1.3. Language Modelling

Our approach is to use a constrained language model (LM) to restrict the search space. In total 39 LMs were generated based on the responses to each of the 39 questions. These responses were manually transcribed at the orthographic level. Filled pauses, restarts, and repetitions were also annotated.

Filled pauses are common in everyday spontaneous speech and generally do not hamper communication. It seems therefore that students using a CALL application should be allowed to produce a limited amount of filled pauses. In our material 46% of the utterances contain one or more filled pauses and almost 13% of all transcribed units are filled pauses.

However, 11% of the utterances contain one or more other disfluencies such as restarts, repairs and repetitions. While these also occur in normal speech, albeit less frequently, we think that in a CALL application for training oral proficiency students should be stimulated to produce fluent speech. On these grounds, we decided not to tolerate restarts, repetitions and repairs and to ask the students to try again when one of these phenomena is produced. Therefore, in our research we did not focus on restarts, repairs, and repetitions, we only included their orthographic transcriptions in the LM and their manual phonetic transcriptions in the lexicon.

The LMs are implemented as FSMs with parallel paths of orthographic transcriptions of every unique answer to the question. A priori each path is equally likely. An example of such a question is "Hoe wilt u naar deze stad reizen?" (How do you want to travel to this city?) and a small part of the responses is

1) /ik gaat met de vliegtuig/ (/I am going by plane/*),

2) /ik ga met detrein/ (/I am going by train/),

3) /met devliegtuig/ (/by plane/*),

4) /met hetvliegtuig/ (/by plane/).

The baseline LM that is generated from this list is depicted in Figure 2. Each of the parallel paths with words on the arcs represents a unique answer to a question. Silence is possible before and after each word (not shown).

To be able to decode possible filled pauses between words, we generated another LM with self-loops added in every node. Filled pauses are represented in the pronunciation lexicon as /@/ or /@m/, phonetic representations of the two most common filled pauses in Dutch. The filled pause loop penalty was empirically optimized. An example of this language model is depicted in Figure 3.

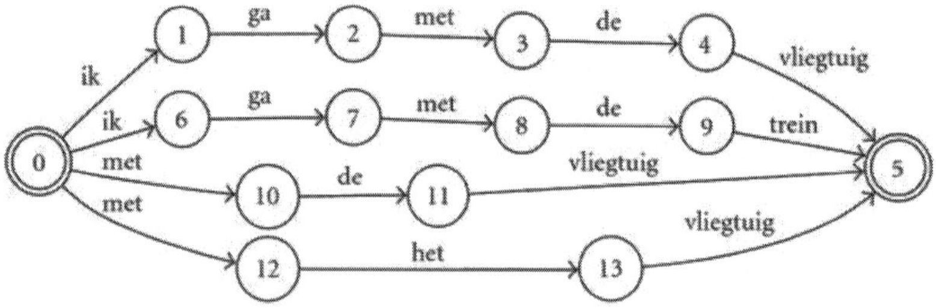

Figure 2. Baseline language model.

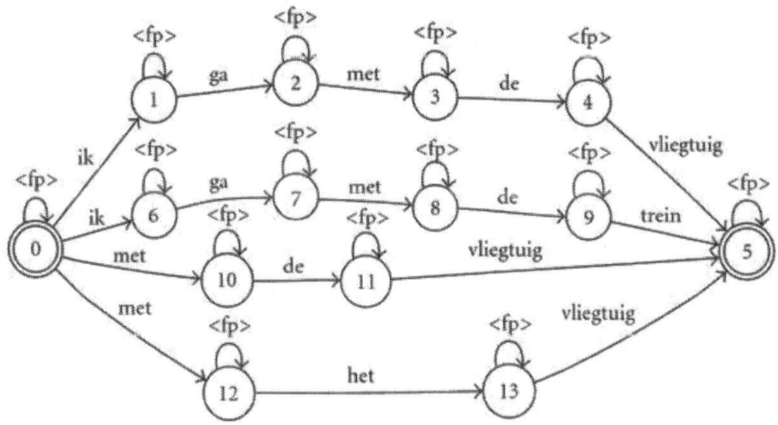

Figure 3. Language model with filled pause loops.

To examine whether filled pause loops are an adequate way of modelling filled pauses, we also experimented with an oracle LM. This is an LM containing the reference orthographic transcriptions, which include the manually annotated filled pauses without filled pause loops.

4.1.4. Acoustic Modelling

We trained three-state tied Gaussian Mixture Models (GMM). Baseline triphone models were trained on 42 hours of native read speech from the CGN corpus [47]. In total 11 660 triphones were created, using 32 738 Gaussians.

As discussed in Section 2.1, it has been observed in several studies that by adapting or retraining native acoustic models (AM) with non-native speech, decoding performance can be increased. To investigate whether this is also the case in a constrained task as described in this paper, we retrained the baseline acoustic models with non-native speech.

New AMs were obtained by doing a one-pass Viterbi training based on the native AMs with 6 hours of non-native read speech from the JASMIN corpus. These utterances were spoken by the same speakers as those in our test material (comparable to an enrollment phase).

Triphone AMs are the de facto choice for most researchers in speech technology. However, the expected performance gain from modelling context dependency by using triphones over monophones might be minimal in a constrained task. Therefore, we also experimented with non-native monophone AMs trained on the same non-native read speech.

4.1.5. Lexical Modelling

The baseline pronunciation lexicon contains canonical phonemic representations extracted from the CGN lexicon. The distribution of sizes of the 39 lexicons is depicted in Figure 4.

As explained in Section 2.1 non-native pronunciation generally deviates from native pronunciation, both at the phonetic and the phonemic level. To model pronunciation variation at the phonemic level, we added pronunciation variants to the lexicon.

To derive pronunciation variants, we extracted context-dependent rewrite rules from an alignment of canonical and realized phonemic representations of non-native speech from the JASMIN corpus (the test material was excluded). Prior probabilities of these rules were estimated by taking the relative frequency of rule applications in their context.

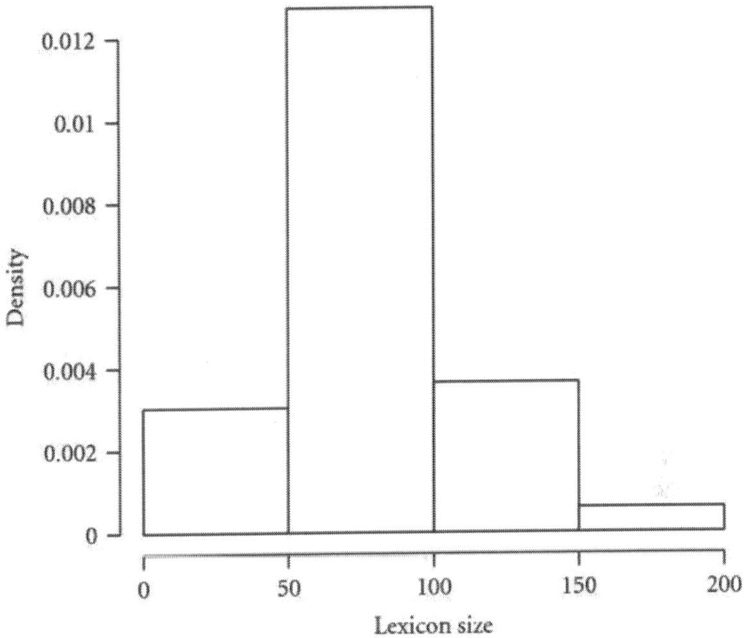

Figure 4. Distribution of lexicon sizes.

We generated pronunciation variants by successively applying the derived rewrite rules to the canonical representations in the baseline lexicon. Variant probabilities were calculated by multiplying the applied rule probabilities. Canonical representations have a standard probability of one. Afterwards, probabilities of pronunciation variants per word were normalized so that these probabilities sum to one.

By introducing a cutoff probability, pronunciation lexicons were created that contain only variants above this cutoff. In this way lexicons with on average 2, 3, 4, and 5 variants per word were created.

4.1.6. Evaluation

We evaluated the speech decoding setups using the utterance error rate (UER), which is the percentage of utterances where the 1-Best decoding result deviates from the transcription. Filled pauses are not taken into account during evaluation. That is, decoding results and reference transcriptions were compared after deletion of filled pauses. For each UER the 95% confidence interval was calculated to evaluate whether UERs between conditions were significantly different.

As explained in the introduction, we do not expect our method to carry out a detailed phonetic analysis in the first phase. Since it is not necessary to discriminate between phonetically close responses at this stage, a decoding result can be classified as correct when its phonetic distance to the corresponding transcription is below a threshold. The phonetic distance was calculated through an alignment program that uses a dynamic programming algorithm to align transcriptions on the basis of distance measures between phonemes represented as combinations of phonetic features [48]. These phonemic transcriptions were made using the canonical pronunciation variants from the words in the orthographic transcriptions.

4.2. Results

In Table 1, the UERs for the different language models and acoustic models can be observed. In all cases, the LM with filled pause loops performed significantly better than the LM without loops. Furthermore, the oracle LM with manually annotated filled pauses (with positions) did not perform significantly better than the LM with loops.

Decoding setups with AMs retrained on non-native speech performed significantly better than those with AMs trained on native speech. The performance difference between monophone and triphone AMs was not significant.

As expected, error rates are lower when evaluating using clusters of phonetically similar responses. To better appreciate the results in Table 1 it is important to get an idea of the meaning of these distances. The distances between the example responses in Section 4.1.3 are shown in Table 2. The density of the phonetic distances between all response pairs to all questions is depicted in Figure 5. Since there are only few responses with a phonetic distance smaller than 5, differences between 0 and 5 are marginal. Performance differences between 0 (equal to transcription) and

10 (one of the answers with a phonetic distance of 10 or smaller to the 1-Best equals the transcription) and between 5 and 15 were significant.

Table 1. This table shows the UERs for the different language models: without FP loops, with FP loops and with FP positions, and different acoustic models: trained on native speech (triphone) and retrained on non-native speech (triphone and monophone). All setups used the baseline canonical lexicon. The columns 0, 5, 10, 15 indicate at what phonetic distance to the reference transcription the decoding result is classified as correct.

AM	LM	0	5	10	15
Native (tri)	without loops	28.9	28.4	26.1	24.6
Native (tri)	with loops	14.9	14.6	12.6	11.0
Native (tri)	with positions	14.7	14.4	13.1	12.0
Non-native(tri)	without loops	22.4	22.0	19.9	18.4
Non-native(tri)	with loops	10.0	9.7	7.9	6.9
Non-native(tri)	with positions	9.4	9.1	7.8	7.1
Non-native(mono)	with loops	11.9	11.5	9.3	8.1

Table 2. Phonetic distances between the example responses: (1) "ik gaat met de vliegtuig", (2) "ik ga met de trein", (3) "met de vliegtuig", (4) "met het vliegtuig".

Response	1	2	3	4
1	0.0	—	—	—
2	20.5	0.0	—	—
3	15.0	23.5	0.0	—
4	23.5	30.0	10.0	0.0

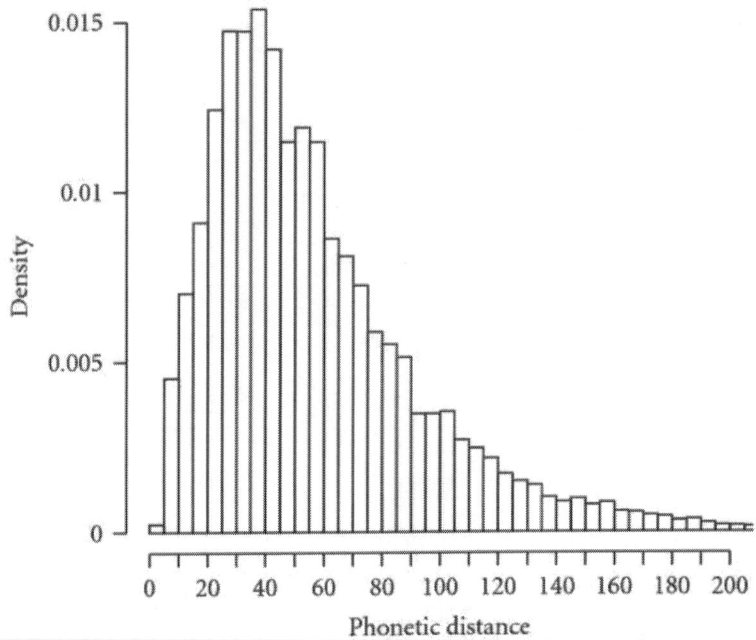

Figure 5. The distribution of phonetic distances between all response pairs to all questions.

As can be seen in Table 3, performance decreased using lexicons with pronunciation variants generated using data-driven methods. The more variants are added, the worse the performance. Furthermore, there is no significant difference between using equal priors or estimated priors.

4.3. Discussion

The results presented in the previous section indicate that large and significant improvements could be obtained by optimizing the language model and the acoustic models. On the other hand, pronunciation modelling at the level of the lexicon did not produce significant improvements. On the contrary, adding variants to the lexicon caused a decrease in performance. Adding estimated prior probabilities to the variants improved the results somewhat, but still the error rates remain higher than those for the canonical lexicon. These results might be surprising because, in general, adding a limited number of carefully selected pronunciation variants to the lexicon helps improve performance to a certain extent [29, 30]. However, in the case of non-native speech this strategy is not always successful [31]. Possible explanations might be sought in the nature of the variation that characterizes non-native speech. Non-native speakers are likely to replace target language phonemes by phonemes from their mother tongue [3, 5]. When the non-native speech is heterogeneous in the sense that it is produced by speakers with different mother tongues, as in our case, it may be extremely difficult to capture the rather diffuse pattern of variation by including variants in the lexicon (see also [4]).

Table 3. UERs for different lexicons: canonical, 2–5 variants with and without priors. These rates are obtained by using non-native triphone acoustic models and language models with filled pause loops.

Lex	Priors	0	5	10	15
canonical	—	10.0	9.7	7.9	6.9
2 var	No	10.0	9.9	8.2	6.7
2 var	Yes	10.0	9.7	8.3	7.0
3 var	No	11.2	10.9	8.5	7.1
3 var	Yes	10.6	10.1	8.7	7.2
4 var	No	11.5	11.3	8.9	7.5
4 var	Yes	10.4	10.9	9.7	7.2
5 var	No	11.5	11.3	8.9	7.5
5 var	Yes	10.4	10.0	8.7	7.2

The findings that better results are obtained with non-native acoustic models and with a language model with filled pause loops are not surprising, after all the utterances are spoken by non-natives, recorded in the same environment and contain a lot of filled pauses. In fact, these results do not differ significantly from the results obtained with an oracle language model, in which the exact position of the filled pauses is copied from the manual transcriptions. This is an important result because non-natives are known to produce numerous filled pauses in unprepared, extemporaneous speech [12]. From these results we can conclude that external filled pause detection, for which better results were found for a large vocabulary task [49], is not necessary in this case.

Another reassuring result is that performance improved using non-native acoustic models. These were obtained by retraining native models on a relatively small amount (around 8 minutes per speaker) of non-native read speech material. It appears that this was sufficient to obtain significantly better results. In the final application we might then use a relatively short enrolment phase and do acoustic model retraining (and/or online speaker adaptation), to obtain better recognition results.

While in this experiment the correct transcription of the response was always in the language model, our system must also be able to reject utterances when they are not present in the language model, while still accepting correctly recognized utterances. This is the topic of the experiment presented in the following section.

5. EXPERIMENT 2: UTTERANCE VERIFICATION

The goal of this experiment is to develop a procedure for utterance verication. Our approach consists of combining an acoustic likelihood ratio with duration-related predictors into one confidence measure.

5.1. Method

5.1.1. Material

We used the same material as in the first experiment, but to simulate the case in which the spoken utterance is not present in the list, we also generated language models in which the correct utterance is left out. In this way, each of the 1325 utterances in our dataset is decoded two times: one time when its representation is present in the language model and one time when it is not present.

5.1.2. Confidence Predictors

As mentioned in Section 4.2, posterior probability estimation using rich word lattices is often used in large vocabulary applications, where it usually provides accurate confidence measures, although it is computationally expensive. Since in our case the search space only contains a limited set of sequences of words, the decoding lattice is not rich enough to estimate $p(\mathbf{x})$ (see (4)). Estimating $p(\mathbf{x})$ on the basis of a free phone recognizer (FPR) is a more simple and faster approach, generally giving reasonably good results. For these reasons, we have used the ratio:

$$\frac{p(\mathbf{x} \mid \hat{W})p(\hat{W})}{p(\mathbf{x} \mid u_{\mathrm{FPR}})p(u_{\mathrm{FPR}})} \qquad (7)$$

as our baseline confidence measure. However, because we have equal prior probabilities for all language model paths and we do not use a language model during free phone recognition the priors $p(\hat{W})$ and $p(u_{\mathrm{FPR}})$ can be discarded and (7) boils down to:

$$\mathrm{LR} = \frac{p(\mathbf{x} \mid \hat{W})}{p(\mathbf{x} \mid u_{\mathrm{FPR}})} \quad . \qquad (8)$$

This ratio bears a close relation to (6) used in the statistical hypothesis testing approach to UV. The main difference is that in the denominator in (8) all paths are used, while in (6) only the alternative paths are used to compare with the recognition result to be verified. Modelling the alternative paths in an antimodel is especially difficult in our task because it is very difficult to determine what exactly it should represent if the utterance is produced by language learners with generally low levels of proficiency and very diverse L1 backgrounds (see also [44]). Furthermore, training such an antimodel requires a large amount of non-native speech data that is not available for Dutch.

We hypothesize that combining our baseline CM (LR) with other predictors that contain additional information about the quality of the recognition result will give better results than using LR alone. However, using the average hypothesis density in the word lattice as a predictor is probably not informative because in our task the word lattice is very small and contains very few competing hypotheses. Furthermore, a predictor like acoustic stability is difficult to define because different weightings of the language model have no effect on the combination score (because a priori each sequence of words in the language model is equally likely).

We expect that phone durations might contain additional information, because the phone segmentation of an incorrectly decoded sequence of words will generally be characterized by deviations in phone durations and this is not directly coded in the acoustic likelihoods in LR. Therefore, we want to add information about these phone duration deviations.

When the input speech representation is not present in the list and the utterance is recognized as another sequence of words that is present in the LM, the phone segmentation of this sequence of words will generally be characterized by deviations in phone durations. A straightforward way to capture this is to count the phones in the segmentation with durations that deviate substantially from the mean phone duration. We have implemented this by using predictors similar to those introduced in [42].

Phone duration distributions were derived from manually verified phonemic transcriptions of 42 hours of read native speech from the CGN corpus [47]. For each of the 46 phonemes the 1st, 5th, 95th, and 99th percentile duration was calculated from these distributions. The predictors that were extracted from the segmentation are the number of phonemes in the decoded utterance that are shorter than the 1st (*nr_shorter_1*) and 5th (*nr_shorter_5*) percentile and the number of phonemes that are longer than the 95th (*nr_longer_95*) and 99th (*nr_longer_99*) percentile durations. These predictors were normalized by the total number of phonemes in the recognized utterance.

5.1.3. Predictor Combination

To combine the five predictors, that is, *LR*, *nr_shorter_1*, *nr_shorter_5*, *nr_longer_95*, *nr_longer_99*, into one confidence measure we have used a logistic regression model. Logistic regression modelling is a straightforward and fast method known to produce accurate predictions when a binary variable is a linear function of several explanatory variables [50]. It fits the logit of the probability (logarithm of the odds) of a binary event as a linear function of the set of explanatory variables:

$$\text{logit}(p(y \mid \mathbf{p})) = \frac{p(y \mid \mathbf{p})}{1-p(y \mid \mathbf{p})} = \beta_0 + \sum_{i=1}^{N} \beta_i x_i \,, \qquad (9)$$

where $p(y \mid \mathbf{p})$ is the probability of a correctly or incorrectly decoded utterance y given the confidence predicting variables $\mathbf{P}$. The optimal weights β are chosen through Maximum Likelihood Estimation (MLE) in WEKA [51]. We trained and tested the model by using Leave- One-Speaker-Out crossvalidation where the model is trained on all speakers except one and then tested on the utterances of the speaker that were left out during training. This is repeated until all speakers are tested.

5.1.4. Evaluation

We evaluated the discriminative ability of our utterance verifier using Receiver Operator Characteristic (ROC) curves, in which the two types of error rates, that is, the false-positive and false-negative rates, are plotted for different thresholds. Using the point on the ROC curve where the error rates of both types are equal, the equal error rate (EER), the different confidence indicators and their combinations are evaluated. 95% confidence intervals were calculated to investigate whether differences between EERs were significantly different.

5.2. Results

The utterance error rate (UER) of our speech decoder on the set of decoding results where the correct transcription was present in the LM was 10.0% (see Section 4.2). In this case errors consist of substitutions with competing language model paths. The UER on the set without the correct transcriptions in the LM was of course 100.0%, so on average 55.0% of all the cases was incorrectly recognized.

The task for the UV was to discriminate the correctly and incorrectly recognized cases. In Table 4, this ability is shown in terms of EER for the individual predictors and several predictor combinations. ROC curves of the best performing predictor and two combinations are shown in Figure 6.

Within the individual predictors LR performs best (14.4%) and all the duration-related predictors perform much worse. The best result for a single duration predictor is 27.3% for *nr_shorter_1*. When we combined all duration-related predictors, *duration_comb*, the EER relative to the best performing duration-related predictor dropped significantly from 27.3%

(with a confidence interval ± 1.7) to 25.3%. Finally, by combining the LR with*duration_comb*, the EER relative to LR decreased significantly by 4.1% from 14.4% to 10.3%.

Table 4 Equal error rates (EER) for the individual features LR, *nr_shorter_1, nr_shorter_5, nr_longer_95, nr_longer_99* and the combinations *duration_comb* (*nr_shorter_1, nr_shorter_5, nr_longer_95, nr_longer_99*) and all features, *all*.

Features	EER
LR	14.4%
nr_shorter_1	27.3%
nr_shorter_5	27.4%
nr_longer_95	35.8%
nr_longer_99	38.5%
duration_comb	25.3%
all	10.3%

In Tables 5(a) and 5(b), percentages are shown using the EER threshold and using all predictors for the two different sets of decoding results, with and without the correct transcription in the LM, respectively. For example, in the set of results with the correct transcription in the LM, 80.8% is classified as correct when it indeed was correctly decoded and 9.2% was classified as incorrect (false reject). In the set without the correct transcription in the LM 91.7% was classified as incorrect when it was incorrectly decoded, and 8.3% was classified as correct (false accept). The performance on the whole dataset is shown in Table 5(c).

5.3. Discussion

The duration-related predictors have a weak performance individually, but they still contain additional information relative to the likelihood ratio LR. The duration-related predictor distributions of correctly and incorrectly decoded utterances overlap severely. This was still the case when we normalized these predictors for the speaking rate within the utterance or when we used the probability of the phoneme durations in the utterance as a predictor. The latter we calculated through a kernel density estimation of the duration probability density per phoneme trained on the CGN native read speech data. Using these more complex predictors the model was not able to make substantially better predictions.

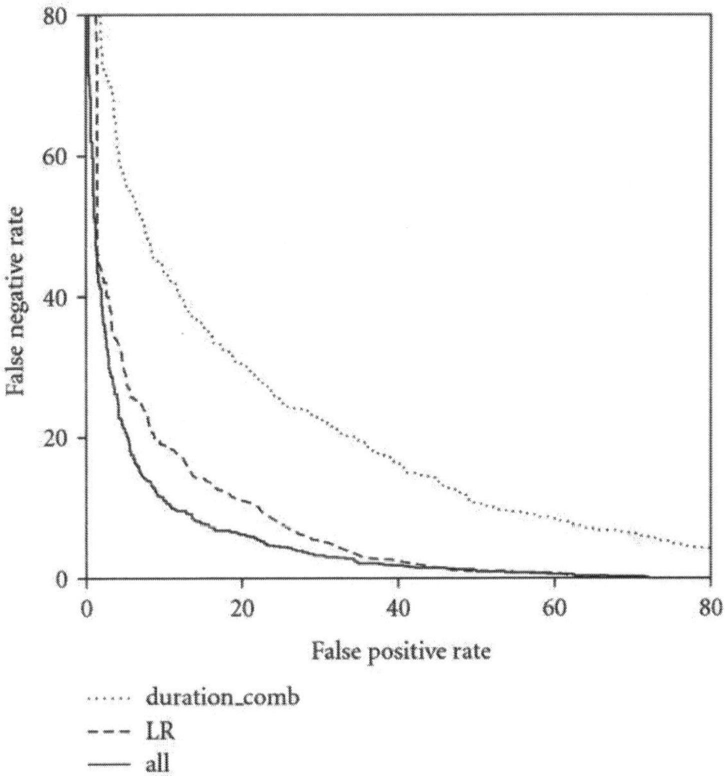

Figure 6. ROC curves for the feature LR and the combinations ***duration_comb*** and ***all*** .

By introducing a UV procedure and using the EER threshold, we are able to filter out 91.7% of the utterances that are not in the predicted list of responses. This comes with the cost of also rejecting utterances that are correctly decoded and accepting utterances that are incorrectly decoded. The ratio between these error rates depends on the threshold setting. We will discuss threshold calibration in the following section.

Table 5. Percentages of correctly and incorrectly classified decoding results of the two different subsets and the total set using the global EER threshold and all predictors. (a) Percentages of decoding result classification on the set where the correct transcription was in the language model. (b) Percentages of decoding result classification on the set where the correct transcription was not present in the language model. (c) Percentages of decoding result classification on the whole dataset.

(a)

		Actual	
		Correct	Incorrect
Predicted	Correct	80.8%	3.0%
	Incorrect	9.2%	7.0%

(b)

		Actual	
		Correct	Incorrect
Predicted	Correct	—	8.3%
	Incorrect	—	91.7%

(c)

		Actual	
		Correct	Incorrect
Predicted	Correct	40.4%	5.6%
	Incorrect	4.6%	49.4%

6. GENERAL DISCUSSION

We carried out two experiments in order to evaluate methods for utterance selection and utterance verification which are going to be used in a CALL application for low-proficient L2 learners of Dutch. For utterance selection with the transcription of the response in the language model, our best error rates were between 10.0% and 6.9% after optimizing acoustic and language models. In 90% of the cases, the decoding result was equal to the corresponding transcription of the response (phonetic distance of 0) and in 93.1% of the cases, the decoder was able to select a cluster of transcriptions with a phonetic distance of 15 or smaller to the 1-Best in which the corresponding transcription was present.

Using an utterance verifier that combined acoustic likelihoods and duration information of the decoding result, 89.8% of the correctly decoded responses is accepted and 70% of the incorrectly decoded utterances could be rejected when the transcription of the response was present in the language model. In addition, 91.7% of the utterances with no representation in the language model could correctly be rejected.

These results apply when we only perform error detection to the 1-Best decoding result, but as explained in Section 3 error detection will probably be performed on the cluster of responses that have a small phonetic distance to the 1-Best decoding result. For example, if it is not clear whether a segment or a (short) word was pronounced or not, this can be ascertained in the second step through a more detailed analysis [19]. At the moment, we think that in the second step we can handle utterances with a phonetic distance smaller than 5, which usually corresponds to a difference of 1 or 2 segments, or possibly even utterances with a phonetic distance smaller than 10, which often boils down to a deviation by a short word. For the latter category, the best result obtained is an error rate of around 8%. This is encouraging, especially if we keep in mind that in a language learning application we can be conservative, in the sense that if we are not sufficiently confident about the recognition result we can always ask the language learner to try again.

Until now we have evaluated the performance of UV using the EER threshold, but this might not be the optimal threshold setting in the actual application. In our application the recognized utterance will be probably shown to the user so that he/she knows whether the utterance was correctly recognized, and where the feedback is based on. If the system makes an error in recognizing the utterance, this will then be clear for the user. The system can make two types of errors: (a) a false rejection, in which case a correctly decoded utterance is classified as incorrect by the UV or (b) a false acceptance, in which case an incorrectly decoded utterance is classified as correct. To determine which of these errors is more detrimental at this stage of the application, it is necessary to consider how such errors can be handled in the application and what their possible consequences are. In the case of a rejection, and therefore also of a false rejection, it is possible to ask the user to repeat the utterance. In concrete terms then, a false rejection implies that the user is unnecessarily asked to repeat the utterance. In the case of a false acceptance an utterance will be shown to the user that (s)he actually did not produce. This type of error would seem to be more detrimental because it can affect the credibility of the system.

However, the degree of seriousness will depend on the degree of discrepancy between the utterance that was actually produced and the one that was recognized and shown by the system: the larger the deviation the more serious the error. On the other hand, large deviations are less likely than small deviations. On the basis of such considerations we can indicate the seriousness of the two types of errors and therefore the costs that should be assigned to false rejections and false acceptances.

There are now three different factors that are important in choosing an application-dependent threshold, namely (1) the prior probability of a correct decoding $P_{correct}$, (2) the cost of a false rejection C_{FR}, and (3) the cost of a false acceptance C_{FA} . To formalize the idea of taking into account different error costs and different prior distributions in the process of choosing a threshold, we can estimate the total cost of a specific threshold setting with a cost function:

$$C_{total} = P_{FR} C_{FR} P_{correct} + P_{FA} C_{FA} (1 - P_{correct}) , \qquad (10)$$

where P_{FR} and P_{FA} are the probabilities of false rejection and false acceptance, respectively. This kind of cost function is also used in the NIST evaluation of speaker recognition systems [52]. Minimizing C_{total} on a development set will provide us with the optimal threshold setting given the application-dependent parameters C_{FR} , C_{FA}, and $P_{correct}$. Using the UV with this application-dependent threshold calibration procedure could make an excellent research vehicle for future experiments with different error costs.

ACKNOWLEDGMENT

The DISCO project is carried out within the STEVIN programme which is funded by the Dutch and Flemish Governments (http://taalunieversum.org/taal/technologie/stevin/).

REFERENCES

1. Van Compernolle D:**Recognizing speech of goats, wolves, sheep and ... non-natives.** *Speech Communication* 2001,**35**(1-2):71-79.

2. Tomokiyo L: *Recognizing non-native speech: characterizing and adapting to non-native usage in speech recognition, Ph.D. dissertation*. Carnergie Mellon University; 2001.

3. Bouselmi G, Fohr D, Illina I, Haton J-P: **Multilingual non-native speech recognition using phonetic confusion-based acoustic model modification and graphemic constraints.** *Proceedings of the 9th International Conference on Spoken Language Processing (ICSLP '06), 2006* **1:** 109-112.

4. Benzeghiba M, De Mori R, Deroo O, *et al.*: **Automatic speech recognition and speech variability: a review.** *Speech Communication* 2007, **49**(10-11):763-786.

5. Flege JE, Munro MJ, Mackay IRA: **Effects of age of second-language learning on the production of English consonants.** *Speech Communication* 1995, **16**(1):1-26.

6. Bohn O-S, Flege J: **The production of new and similar vowels by adult German learners of English.** *Studies in Second Language Acquisition* 1992, **14**(2):131-158.

7. Bradlow AR, Pisoni DB, Akahane-Yamada R, Tohkura Y: **Training Japanese listeners to identify English /r/and /l/: IV. Some effects of perceptual learning on speech production.** *Journal of the Acoustical Society of America* 1997,**101**(4):2299-2310.

8. Neri A, Cucchiarini C, Strik H: **Selecting segmental errors in L2 Dutch for optimal pronunciation training.** *International Review of Applied Linguistics in Language Teaching* 2006, **44**(4):357-404.

9. DeKeyser RM: **What makes learning second-language grammar difficult? A review of issues.** *Language Learning*2005, **55**(supplement 1):1-25.

10. Cucchiarini C, Strik H, Boves L: **Different aspects of expert pronunciation quality ratings and their relation to scores produced by speech recognition algorithms.** *Speech Communication* 2000, **30**(2-3):109-119.

11. Cucchiarini C, Strik H, Boves L: **Quantitative assessment of second language learners' fluency by means of automatic speech recognition technology.** *Journal of the Acoustical Society of America* 2000, **107**(2):989-999.

12. Cucchiarini C, Strik H, Boves L: **Quantitative assessment of second language learners' fluency: comparisons between read and spontaneous speech.** *Journal of the Acoustical Society of America* 2002, **111**(6):2862-2873.

13. Moehle D: **A comparison of the second language speech production of different native speakers.** In *Second Language Productions*. Narr, Tübingen, Germany; 1984.

14. Towell R, Hawkins R, Bazergui N: **The development of fluency in advanced learners of French.** *Applied Linguistics*1996, **17**(1):84-119. 10.1093/applin/17.1.84

15. Ehsani F, Bernstein J, Najmi A: **Interactive dialog system for learning Japanese.** *Speech Communication* 2000,**30**(2):167-177.

16. Raux A, Eskenazi M: **Using task-oriented spoken dialogue systems for language learning: potential, practical applications and challenges.** *Proceedings of the InSTIL/ICALL Symposium on Computer Assisted Learning, June 2004, Venice, Italy*

17. Bortfeld H, Leon SD, Bloom JE, Schober MF, Brennan SE: **Disfluency rates in conversation: effects of age, relationship, topic, role, and gender.** *Language and Speech* 2001, **44**(2):123-147.

18. Menzel W, Herron D, Bonaventura P, Morton R: **Automatic detection and correction of non-native English pronunciations.** *Proceedings of the International Workshop on Integrating Speech Technology in the Language Learning and Assistive Interface (InStil '00), 2000* 49-56.

19. Cucchiarini C, Neri A, Strik H: **Oral proficiency training in Dutch L2: the contribution of ASR-based corrective feedback.** *Speech Communication* 2009, **51**(10):853-863.

20. Byrne W, Knodt E, Khudanpur S, Bernstein J: **Is automatic speech recognition ready for non-native speech? A data collection effort and initial experiments in modeling conversational Hispanic English.** *Proceedings of the Speech Technology in Language Learning (STiLL '98), May 1998, Stockholm, Sweden* 37-40.

21. Gerosa M, Giuliani D: **Preliminary investigations in automatic recognition of English sentences uttered by Italian children.** *Proceedings of the InSTIL/ICALL Symposium on Computer Assisted Learning (ICALL '04), 2004* 9-13.

22. Neumeyer L, Franco H, Weintraub M, Price P: **Automatic text-independent pronunciation scoring of foreign language student speech.** *Proceedings of the International Conference on Spoken Language Processing (ICSLP '96), 1996* **3:** 1457-1460.

23. Franco H, Neumeyer L, Digalakis V, Ronen O: **Combination of machine scores for automatic grading of pronunciation quality.** *Speech Communication* 2000, **30**(2):121-130.

24. Deville G, Deroo O, Gielen S, Leich H, Vanparys J: **Automatic detection and correction of pronunciation errors for foreign language learners the Demosthenes application.** *Proceedings of the 6th European Conference on Speech Communication and Technology (EUROSPEECH '99), September 1999, Budapest, Hungary* **2:** 843-846.

25. Kawai G, Hirose K: **A method for measuring the intelligibility and nonnativeness of phone quality in foreign language pronunciation training.** *Proceedings of the 5th International Conference on Spoken Language Processing (ICSLP '98), November-December 1998, Sydney, Australia* 1823-1826.

26. Witt S: *Use of speech recognition in computer assisted language learning, Ph.D. dissertation*. University of Cambridge, Cambridge, UK; 1999.

27. Witt S, Young S: **Off-line acoustic modelling of non-native accents.** *Proceedings of the 6th European Conference on Speech Communication and Technology (EUROSPEECH '99), September 1999, Budapest, Hungary* 1367-1370.

28. Tomokiyo L: **Handling non-native speech in LVCSR: a preliminary study.** *Proceedings of the Incorporating Speech Technology in Language Learning (InSTIL '00), 2000* 62-63.

29. Strik H, Cucchiarini C: **Modeling pronunciation variation for ASR: a survey of the literature.** *Speech Communication*1999, **29**(2–4):225-246.

30. Strik H: **Pronunciation adaptation at the lexical level.** In *Proceedings of the ISCA Tutorial & Research Workshop Adaptation Methods for Speech Recognition (ITRW '01), 2001, Sophia Antipolis, France* Edited by: Juncqua J-C, Wellekens C.

31. Goronzy S, Rapp S, Kompe R: **Generating non-native pronunciation variants for lexicon adaptation.** *Speech Communication* 2004, **42**(1):109-123.

32. Livescu K, Glass J: **Lexical modeling of non-native speech for automatic speech recognition.** *Proceedings of the IEEE International Conference on Acoustics, Speech and Signal Processing (ICASSP '00), June 2000, Istanbul, Turkey* **3:**1683-1686.

33. Markowitz J: *Using Speech Recognition*. Prentice-Hall, Englewood Cliffs, NJ, USA; 1996.

34. Ehsani F, Knodt E: **Speech technology in computer-aided language learning: strengths and limitations of a new call paradigm.** *Language Learning and Technology* 1998, **2**(1):45-60.

35. Atwell E, Herron D, Howarth P, Morton R, Wick H: **Pronunciation training: requirements and solutions.** ISLE Deliverable 1.4, 1999

36. Waltje J: **CALICO software review—tell me more-German.** *CALICO Journal* 2002, **19**(2):405-418.

37. Jiang H: **Confidence measures for speech recognition: a survey.** *Speech Communication* 2005, **45**(4):455-470.

38. Wessel F, Schlüter R, Macherey K, Ney H: **Confidence measures for large vocabulary continuous speech recognition.** *IEEE Transactions on Speech and Audio Processing* 2001, **9**(3):288-298.

39. Young S: **Detecting misrecognitions and out-of-vocabulary words.** *Proceedings of the International Conference on Acoustics, Speech and Signal Processing (ICASSP '94), 1994*

40. Bouwman G, Boves L: **Utterance verification based on the likelihood distance to alternative paths.** *Proceedings of the 5th International Conference on Text, Speech and Dialogue (TSD '02), September 2002* 213-220.

41. Schaaf T, Kemp T: **Confidence measures for spontaneous speech recognition.** *Proceedings of the IEEE International Conference on Acoustics, Speech and Signal Processing (ICASSP '97), April 1997, Munich, Germany* **2:** 875-878.

42. Goronzy S, Marasek K, Kompe R, Haag A: **Prosodically motivated features for confidence measures.** *Proceedings of the Automatic Speech Recognition: Challenges for the New Millenium (ASR '00), September 2000, Paris, France* **1:** 207-212.

43. Desmukh O, Verma A: **Acoustic-phonetic approach for automatic evaluation of spoken grammar.** *Proceedings of the 9th Annual Conference of the International Speech Communication Association (INTERSPEECH '08), September 2008, Brisbane, Australia* 2614-2617.

44. de Wet F, Cucchiarini C, Strik H, Boves L: **Using likelihood ratios to perform utterance verification in automatic pronunciation assessment.** *Proceedings of the 6th European Conference on Speech Communication and Technology (EUROSPEECH '99), September 1999, Budapest, Hungary* 173-176.

45. Cucchiarini C, Driesen J, Hamme HV, Sanders E: **Recording speech of children, non-natives and elderly people for HLT applications: the JASMIN-CGN corpus.** *Proceedings of the 6th International Language Resources and Evaluation (LREC '08), May 2008, Marrakech, Morocco*

46. Demuynck K, Roelens J, Compernolle DV, Wambacq P: **SPRAAK: an open source SPeech Recognition and Automatic Annotation Kit.** *Proceedings of the International Conference on Spoken Language Processing (ICSLP '08), September 2008, Brisbane, Australia* 495.

47. Oostdijk N: **The design of the spoken Dutch corpus.** In *New Frontiers of Corpus Research*. Edited by: Peters P, Collins P, Smith A. Rodopi; 2002:105-112.

48. Cucchiarini C: **Assessing transcription agreement: methodological aspects.** *Clinical Linguistics and Phonetics* 1996,**102**(2):131-155.

49. Stouten F, Duchateau J, Martens J-P, Wambacq P: **Coping with disfluencies in spontaneous speech recognition: acoustic detection and linguistic context manipulation.** *Speech Communication* 2006, **48**(11):1590-1606.

50. Mitchell T: *Machine Learning*. McGraw-Hill, New York, NY, USA; 1997.

51. Witten I, Frank E: *Data Mining: Practical Machine Learning Tools and Techniques*. Morgan Kaufmann, San Mateo, Calif, USA; 2005.

52. van Leeuwen DA, Brümmer N: **An introduction to application-independent evaluation of speaker recognition systems.** In *Speaker Classification, Lecture Notes in Computer Science/Artificial Intelligence. Volume 4343*. Edited by: Müller C. Springer, New York, NY, USA; 2007.

Chapter 9

Speech Recognition for Agglutinative Languages

R. Thangarajan[1]

[1] Department of Computer Science and Engineering, Kongu Engineering College, Perundurai, Erode, Tamilnadu, India

1. INTRODUCTION

Speech technology is a broader area comprising many applications like speech recognition, Text to Speech (TTS) Synthesis, speaker identification and verification and language identification. Different applications of speech technology impose different constraints on the problem and these are tackled by different algorithms. In this chapter, the focus is on automatically transcribing speech utterances to text. This process is called Automatic Speech Recognition (ASR). ASR deals with transcribing speech utterances into text of a given language. Even after years of extensive research and development, ASR still remains a challenging field of research. But in the recent years, ASR technology has matured to a level where success rate is higher in certain domains. A well-known example is human-computer interaction where speech is used as an interface along with or without other pointing devices. ASR is fundamentally a statistical problem. Its objective is to find the most likely sequence of words, called hypothesis, for a given sequence of observations. The sequence of observations involves acoustic feature vectors representing the speech utterance. The performance of an ASR system can be measured by aligning the hypothesis with the reference text and by counting errors like deletion, insertion and substitution of words in the hypothesis.

ASR is a subject involving signal processing and feature extraction, acoustics, information theory, linguistics and computer science. Speech signal processing helps in extracting relevant and discriminative information, which is called features, from speech signal in a robust manner. Robustness involves spectral analysis used to characterize time varying properties of speech signal and speech enhancement techniques for making features resilient to noise. Acoustics provides the necessary understanding of the relationship between speech utterances and the physiological processes in speech production and speech perception. Information theory provides the necessary procedures for estimating parameters of statistical models during training phase. Computer science plays a major role in ASR with its implementation of efficient algorithms in software or hardware for decoding speech in real-time.

Currently, an important area in speech recognition is Large Vocabulary Continuous Speech Recognition (LVCSR). A large vocabulary means that system has a vocabulary ranging from 5,000 to 60,000 words. Continuous speech means that the utterances have words which are run together naturally. This is different from isolated word speech recognition where each word is demarcated with initial and final silence. Speech recognition algorithms can also be speaker independent, i.e. they are even able to recognize the speech of new users with whom the system is not exposed. Xuedong et al (2001) present a very good reference on algorithms and techniques in speech processing.

1.1. Issues In Speech Variability

Even though state-of-the-art speech recognition systems cannot match human performance, still they can recognize spoken input accurately but with some constraints. Some of the constraints may be speaker dependency, language dependency, speaking style, and applicability to a particular task and environment. Therefore, building a speech recognizer that could recognize the speech of any speaker, speaking in any language, style, domain and environment is far from realization.

a. Context variability

In any language, words with different meanings have the same phonetic realization. Their usage depends on the context. There is even more context dependency at the phone level. The acoustic realization of a phone is dependent on its neighboring phones. This is because of the physiology of articulators involved in speech production.

b. Style variability

In isolated speech recognition with a small vocabulary, a user pauses between every word while speaking. Thus it is easy to detect the boundary between words and decode them using the silence context. This is not possible in continuous speech recognition. The speaking rate affects the word recognition accuracy. That is, the higher the speaking rate, the higher the WER.

c. Speaker variability

Every speaker's utterance is unique per se. The speech produced by a speaker is dependent on a number of factors, namely vocal tract physiologies, age, sex, dialect, health, education, etc. For speaker independent speech recognition, more than 500 speakers from different age groups, their sex, educational background, and dialect are necessary to build a combined model. The speaker independent system includes a user enrollment process where a new user can train his voice with the system for 30 minutes before using it.

d. Environment variability

Many practical speech recognizers lack robustness against changes in the acoustic environment. This has always been a major limitation of speech based interfaces used in mobile communication devices. The acoustic environment variability is highly unpredictable and it cannot be accounted for during training of models. A mismatch will always occur between the trained speech models and test speech.

1.2. Measure Of Performance

The ultimate measures of success for any speech recognition algorithms are accuracy and robustness. Therefore, it is important to evaluate the performance of such a system. The WER is one of the most widely used measures for accuracy. The system may be tested on sample utterances from the training data for understanding the system and identification of bugs during the development process. This would result in a better performance than what one can get with test data. In addition, a development set can also be used to test the system and also fine-tune its parameters. Finally, the system can be tested on a test set comprising around 500 speech utterances of 5-10 different users in order to reliably estimate accuracy. This test set should be completely new with respect to training and development. There are three types of word recognition errors in speech recognition:

- Substitution (*Subs*): An incorrect word substituted for a correct word.
- Insertion (*Ins*): An extra word added in the recognized sentence.
- Deletion (*Dels*): A correct word omitted in the recognized sentence.

Generally, a hypothesis sentence is aligned with the correct reference sentence. The number of insertions, substitutions and deletions are computed using maximum substring matching. This is implemented using dynamic programming. The WER is computed as shown in equation (1):

$$WER = \frac{Subs+Dels+Ins}{N} \times 100 \tag{1}$$

Other performance measures are speed and memory footprints. The speed is an important factor which quantifies the turn around time of the system once the speech is uttered. It is calculated as shown inequation (2):

$$speed = \frac{time\ taken\ for\ processing}{utterance\ duration} \times real\ time \tag{2}$$

Obviously, the time taken for processing should be shorter than the utterance duration for a quicker response from the system. Memory footprints show the amount of memory required to load the model parameters.

There are a number of well-known factors which affect the accuracy of an ASR system. The prominent factors are those which include variations in context, speakers and noise in the environment. Research in ASR is classified into different types depending on the nature of the problems, like a small or a large vocabulary task, isolated or continuous speech, speaker dependent or independent and robustness to environmental variations. The state-of-the-art speech recognition systems can recognize spoken input accurately with some constraints. The constraints can be speaker dependency, language dependency, speaking style, task or environment. Therefore, building an automatic speech recognizer which can recognize the speech of different speakers, speaking in different languages, with a variety of accent, in any domain and in any ambience environmental background is far from reality.

ASR for languages like English, French and Czech is well matured. A lot of research and development have also been reported for oriental languages like Chinese and Japanese. But in the Indian scenario, ASR is still in its nascent stage due to the inherent agglutinative nature of most of its official languages. Agglutination refers to the extensive morphological inflection in which one can find a one-to-one correspondence between affixes and syntactic categories. This nature results in a large number of words in the dictionary which hinders modeling and training of utterances, and also creates Out-Of-Vocabulary (OOV) words when deployed.

2. SPEECH UNITS

The objective of this chapter is to discuss a few methods to improve the accuracy of ASR systems for agglutinative languages. The language presented here as a case study is Tamil (ISO 639-3 *tam*). Tamil is a Dravidian language spoken predominantly in the state of Tamilnadu in India and in Sri Lanka. It is the official language of the Indian state of Tamilnadu and also has official status in Sri Lanka, Malaysia and Singapore. With more than 77 million speakers, Tamil is one of the widely spoken languages in the world. Tamil language has also been conferred the status of classical language by the government of India.

Currently, there is a growing interest among Indian researchers for building reliable ASR systems for Indian languages like Hindi, Telegu, Bengali and Tamil. Kumar et al (2004) reported the implementation of a Large Vocabulary Continuous Speech Recognition (LVCSR) system for Hindi. Many efforts have been put to build continuous speech recognition systems for Tamil language with a limited and restricted vocabulary (Nayeemulla Khan and Yegnanarayana 2001, Kumar and Foo Say Wei 2003, Saraswathi and Geetha 2004, Plauche et al 2006). Despite repeated efforts, a LVCSR system for the foresaid languages is yet to be explored to a significant level. Keeping agglutination apart, there are other issues to be addressed like aspirated and un-aspirated consonants, and retroflex consonants.

2.1. Phones, Phonemes and Syllables

Before embarking on the concepts, it is always better to review a few terminologies pertaining to linguistics. In any language, there are acoustic properties of speech units and symbolic representation of lexical units. For more information, please refer (Xuedong et al, 2001).

a. Phonemes and phones

From the acoustics point of view, a phoneme is defined as the smallest segmental unit of sound employed to tell apart meaningfully between utterances. A phoneme can be considered as a group of slightly different sounds which are all perceived to have the same function by the speakers of a language or a dialect. An example of a phoneme is the /k/ sound in the words *kit* and *skill*. It is customary to place phonemes between slashes in transcriptions. However, the phoneme */k/* in each of these words is actually pronounced differently i.e. it has different realizations. It is because the articulators which generate the phoneme cannot move from one position to another instantaneously. Each of these different realizations of the phoneme is called a phone or technically an allophone (in transcriptions, a phone is placed inside a square bracket like [*k*]). A phone can also be defined as an instance of a phoneme. In *kit* [*k*] is aspirated while in *skill* [*k*] is un-aspirated. Aspiration is a period of voiceless-ness after a stop closure and before the onset of voicing of the following vowel. Aspiration sounds like a puff of air after the [*k*] and before the vowel. An aspirated phone is represented as [k^h]. In some languages, aspirated and un-aspirated consonants are treated as different phonemes. Hindi, for instance, has four realizations for [*k*] and they are considered as different phonemes. Tamil does not discriminate them and treats them as allophones.

b. Words

Next comes the representation in symbolic form. According to linguistics, word is defined as a sequence of morphemes. The sequence is determined by the morpho-tactics. A morpheme is an independent unit which makes sense in any language. It could refer to the root word or any of the valid prefixes or suffixes. Therefore what is called a word is quite arbitrary and depends on the language in context. In agglutinative languages a word could consists of a root along with its suffixes – a process known as inflectional morphology. Syntax deals with sentence formation using lexical units. Agglutinative languages, on one hand, exhibit inflectional morphology to a higher extent. On the other hand, the syntactic structure is quite simple which enables free-word ordering in a sentence. The English language, which is not agglutinative, has simpler lexical morphology but the complexity of the syntactic structure is significantly higher.

c. Syllables

A syllable is a unit of organization for a sequence of speech sounds. It is composed of three parts: the onset, the nucleus and the coda. A syllable has a hierarchical structure as shown in Figure 1. It can also be expressed in Backus-Naur Form (BNF) as follows:

<syllable> ::= <onset> <rhyme>

<rhyme> ::= <nucleus> <coda>

A vowel forms the nucleus of the syllable while an optional consonant or consonant cluster forms the onset and coda. In some syllables, the onset or the coda will be absent and the syllables may start and/or end with a vowel.

Generally speaking, a syllable is a vowel-like sound together with some of the surrounding consonants that are most closely associated with it. For example, in the word *parsley*, there are two syllables**[pars.ley]** – CVCC and CVC. In the word *tarragon*, there are three syllables **[tar.ra.gon]** – CVC, CV and CVC. The process of segmenting a word into syllables is called *syllabification*. In English, the syllabification is a hard task because there is no agreed upon definition of syllable boundaries. Furthermore, there are some words like *meal*, *hour* and *tire* which can be viewed as containing one syllable or two (Ladefoged 1993).

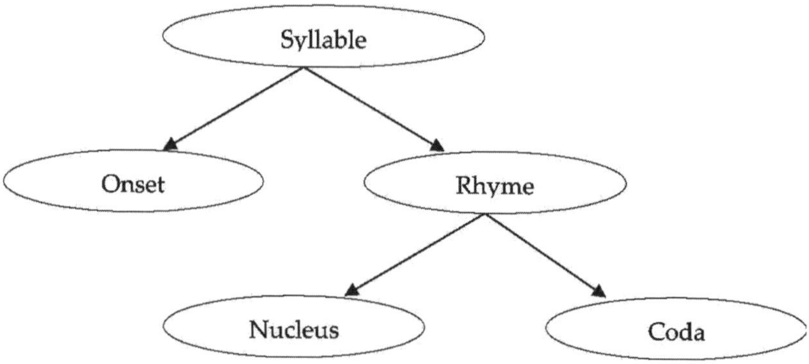

Figure 1. A Syllable's Hierarchical Structure

A syllable is usually a larger unit than a phone, since it may encompass two or more phonemes. There are a few cases where a syllable may only consist of single phoneme. Syllables are often considered the phonological building blocks of words. Syllables have a vital role in a language's rhythm, prosody, poetic meter and stress.

The syllable, as a unit, inherently accounts for the severe contextual effects among its phones as in the case of words. Already it has been observed that a syllable accounts for pronunciation variation more systematically than a phone (Greenberg 1998). Moreover, syllables are intuitive and more stable units than phones and their integrity is firmly based on both the production and perception of speech. This is what sets a syllable apart from a triphone. Several research works using syllable as a speech unit have been successfully carried out for English and other oriental languages like Chinese and Japanese by researchers across the world.

In Japanese language, for instance, the number of distinct syllables is 100, which is very small (Nakagawa et al 1999). However in a language like English, syllables are large in number. In some studies, it is shown that they are of the order of 30,000 syllables in English. The number of lexically attested syllables is of the order of 10,000. When there are a large number of syllables, it becomes difficult to train syllable models for ASR (Ganapathiraju et al 2001).

3. AGGLUTINATIVE LANGUAGES – TAMIL

Tamil language, for instance, employs agglutinative grammar, where suffixes are used to mark noun class, number, and case, verb tense and other grammatical categories. As a result, a large number of inflectional variants for each word exist. The use of suffixes is governed by morpho-tactic rules. Typically, a *STEM* in Tamil may have the following structure.

STEM +negative+participle +nominalization +plural +locative +ablative +inclusive

For each stem, there are at least $2^7 = 128$ inflected word forms, assuming only two affixes of each type. Actually, there may be more than two options, but there may be gaps. In contrast, English has maximally 4 word forms for a verb as in *swim, swims, swam* and *swum*, and for nouns as in *man,man's, men* and *men's*. Hence, for a lexical vocabulary of 1,000, the actual Tamil words list of inflected forms will be of the order of 1,28,000.

3.1. Inflectional Morphology

The Parts of Speech (POS) categories in Tamil take different forms due to inflections. According toRajendran (2004) morphological inflections on nouns include gender and number. Prepositions take either independent or noun combined forms with various cases like accusative, dative, instrumental, sociative, locative, ablative, benefactive, genitive, vocative, clitics and selective. Table 1 arrays a list of examples of cases and their possible suffixes.

The verbs in Tamil take various forms like simple, transitive, intransitive, causative, infinitive, imperative and reportive. Verbs are also formed with a stem and various suffix patterns. Some of the verbal suffix patterns are shown in Table 2. Rajendran et al (2003) had done a detailed study on computational morphology of verbal patterns in Tamil.

Adjectives and adverbs are generally obtained by attaching the suffix – ஆன and ஆக to noun forms respectively. Tamil often uses a verb, an adjective or an adverb as the head of a noun phrase. This process is called nominalization where a noun is produced from another POS with morphological inflections.

Table 1. Case Suffixes used with Noun in Tamil

Cases	Suffixes
Accusative	ஏ, ஐ
Dative	க்கு, ற்கு
Instrumental	ஆல்
Sociative	ஓடு, உடன்
Locative	இல், உள், இடம்
Ablative	இருந்து
Benefactive	க்காக, ற்காக
Genetive	இன், அது, உடைய
Vocative	ஏ
Clitics	உம், ஓ, தான்
Selective	ஆவது
Interrogative	ஆ, ஓ

3.2. Morpho-Phonology

Morpho-phonology, also known as *sandhi*, wherein two consecutive words combine by deletion, insertion or substitution of phonemes at word boundaries to form a new word is very common in Tamil. In English, one can find morpho-phonology to a limited extent. For example, the negative prefix (*in*) when attached to different words, changes according to the first letter of the word.

in + proper → *improper*

in + logical → *illogical*

in + rational → *irrational*

in + mature → *immature*

Table 2. Verbal Suffixes in Tamil

Suffix	Categories	Sub categories	Suffixes
Tense	Present		கிறு, கின்று, ஆனின்று
	Past		த் , ந் , ற் , இன்
	Future		ப் , வ்
Person	First	Singular	ஏன்
		Plural	ஓம்
	Second	Singular	ஆய்
		Plural	ஈர்கள்
		Honorific	ஈர்
	Third	Male Singular	ஆன் ,அன்
		Female Singular	ஆள், அள்
		Common Plural	ஆர்கள்
		Honorific	ஆர் , அர்
		Neutral Singular	அது
		Neutral Plural	அன
Others	Causative		இ
	Verbal Noun Untensed		அல்
	Infinitive		உ
	Imperative	Plural	உங்கள்
		Negative	ஆதே, ஆது
	Passive		படு
	Future	Negative	மாட், இல்லை
	Optative		முடியும், வேண்டும், கூடும், ஆம்
		negative	முடியாது, கூடாது, வேண்டாம்
	Morpho-phonology (*Sandhi*)		ந், க், ம், ச், த்
	Plural		கள்

However in Tamil, use of morpho-phonology is more common among two adjacent words. The last phoneme of the first word and the first phoneme of the following word combine and undergo a transformation. Based on certain phono-tactic rules and context, there may be no transformation, or an insertion of a consonant, or a deletion of a vowel/consonant, or a substitution of a vowel/consonant. The following examples illustrate this phenomenon.

அரசு (*government*) + பணி (*service*) → அரசுப்பணி

ஆபரணம் (*ornament*) + தங்கம் (*gold*) → ஆபரணத்தங்கம்

ஒன்று (*first*) + ஆவது (*selective case suffix*) → ஒன்றாவது

In the first example, there is an insertion of a consonant (ப்) between the two words. The second example shows a substitution of the last consonant (ம்) of the first word by another consonant (த்). In the third example, there is a deletion of the last vowel (உ i.e. று ற் + உ) of the first word and the consonant (ற்) merges with the incoming vowel (ஆ i.e. ற் + ஆ றா) of the second word. As a result of morpho-phonology, two distinct words combine and sound as a single word. In fact, Morpho-phonology has evolved as a result of context dependencies among the phonetic units at the boundary of adjacent words or morphemes.

To a certain extent morpho-phonology is based on phono-tactics. The following rules encompass most of them.

- 'உ' removal rule: This rule states that when one morpheme ends in the vowel 'உ' and the following morpheme starts with a vowel, then the 'உ' would be removed from the combination.
- 'வ்' and 'ய்' addition rule: When one morpheme ends with a vowel from a particular a set of vowels and the morpheme it joins starts with a vowel then the morpheme 'வ்' or 'ய்' would be added to the end of the first morpheme.
- Doubling rule: According to this rule, when one morpheme ends with either 'ண்', 'ன்', 'ம்' or 'ய்' and the next morpheme starts with a vowel then the 'ண்', 'ன்', 'ம்' or 'ய்' is doubled.
- Insertion of க், ச், ட் or ப்: This rule states that when one morpheme ends with a vowel and the next morpheme begins with either க், ச், ட்

or ப் followed by a vowel there is a doubling of the corresponding க், ச், ட் or ப்.

Apart from these rules, there are instances of morpho-phonology based on the context. For example, பழங் கூடை (old basket) and பழக் கூடை (fruit basket). Therefore, modelling morpho-phonology in Tamil is still a challenging issue for research.

3.3. Pronunciation In Tamil

Generally languages structure the utterance of words by giving greater prominence to some constituents than others. This is true in English where one or more syllables standout as more prominent than the rest. This is typically known as word stress. The same is true for higher level prosody in a sentence where one or more syllables may bear sentence stress or accent. Spoken Tamil language has a number of dialects. People from different parts of Tamilnadu state in India speak different accents. Harold Schiffman (2006) is a good reference for studying the grammar of spoken Tamil. As far as formal Tamil language is concerned, it is assumed that there is no stress or accent (Arden 1934, Arokianathan 1981, Soundaraj 2000) at word level and all syllables are pronounced with the same emphasis. However, there are other opinions that the position of stress in the word is by no means fixed to any syllable of individual word (Marthandan 1983). In connected speech, the stress is found more often in the initial syllable (Balasubramaniam 1980). In some studies (Asher and Keane 2005) it is shown that there is a marked reduction in vowel's duration of non-initial syllables compared to initial syllables.

4. NOVEL METHODS FOR IMPROVING ACCURACY OF ASR SYSTEMS

This section concentrates on two novel approaches adopted in ASR systems for agglutinative languages. In the first approach, an enhanced bi-gram or tri-gram morpheme based language model is designed to reduce the vocabulary size and reliably predict the strings in an agglutinative language.

The second approach leverages the syllabic structure of the word and builds a syllable based acoustic model.

4.1. Language Models

A language model is defined as the probability distribution over strings in a language. Frequencies of patterns of words as they occur in any training corpus are recorded as probability distributions. A language model is an indispensable part of ASR and machine translation systems. Language model helps in reducing the decoder's search space and in generating optimal sequence of words. For strict word-order languages like English, statistical language models have been effectively used because trigram or bigram models accurately model short distance relationship in a sentence. Other sophisticated language models also exist like class based model, distance based model and dependency based model. The performance of a statistical language model is measured in terms of perplexity. The perplexity refers to the branching factor in the search graph. If the perplexity is low, better the performance of a language model.

Owing to resource deficiency in text and annotated corpora in Tamil, building reliable statistical language models is a difficult task. Even with the available corpus, the agglutinative nature of the language further deepens the problem. Statistical studies using large corpora are still in the nascent stage. However, Rajendran (2006) has given a review of various recent works carried out in morphological analysis, morphological disambiguation, shallow parsing, POS tagging and syntactic parsing in Tamil.

For instance, let Wi–1Wi-1 and WiWi be two consecutive words in a text. The probabilityP(Wi|Wi–1)P(Wi|Wi-1) is the bi-gram that measures the correlation between the words Wi–1Wi-1 andWiWi. This bi-gram measure is sufficient for modeling strings of words in a language where inflectional morphology is low. However in agglutinative languages, like Tamil, a more minute measure is warranted. This issue has been successfully resolved by Saraswati and Geetha (2007) with their enhanced morpheme based language model. The size of the vocabulary was reduced by decomposing the words into stems and endings. These sub-word units (morphemes) are stored in the vocabulary separately. The enhanced morpheme-based language model is designed and trained on the decomposed corpus. A Tamil text corpus is decomposed into stem and its associated suffixes using an existing Tamil morphological analyzer (Anandan P et al, 2002). The decomposition helps reduce the number of distinct words by around 40% on two different corpora namely News and Politics. The stems and its endings are marked with a special character '#' for stems and '$' for suffixes in order to co-join them back after recognition is done.

A general morpheme based language model is one where the stem and suffixes are treated as independent words. No distinction is made between a stem and a morpheme. Figure 2 depicts the various probability measures involved in a morpheme based language model. The word Wi−1Wi-1 is split into the stem Si−1Si-1and suffixEi−1Ei-1, and the word WiWiis split into the stem SiSi and suffixEiEi.

In this case, the prediction of suffix Ei−1Ei-1 will be based on Si−1Si-1 which is strongly correlated since a stem can have a few suffixes among the possible 7 suffixes. This information is modeled in P(Ei−1|Si−1)P(Ei-1|Si-1)and P(Ei|Si)P(Ei|Si)which can be reliably gathered from a corpus. However, the correlation between stem SiSi and the suffix of the previous word Ei−1Ei-1 is weak, because the suffix bears very little information of the next word in a sentence. But when it comes to stem to stem correlation, the probability P(Si|Si−1)P(Si|Si-1)can be reliably used, since there is contextual information that exists between adjacent words in a sentence and stem is the primary part of a word. In Tamil language there is strong subject-predicate agreement which leads to contextual information between suffixes of words in a sentence. This information is available in P(Ei|Ei−1)P(Ei|Ei-1)

The perplexity and WER are obtained using a Tamil speech recognition system. While figure 3.a portrays the perplexity of the language models, figure 3.b compares the WER of the ASR system employing both language models.

The results has confirmed that the modified morpheme-based trigram language model with Katz back-off smoothing technique is better perplexity and lower WER on two Tamil corpora. The results confirm that the proposed enhanced morpheme-based language model is much better than the word-based language models for agglutinative languages.

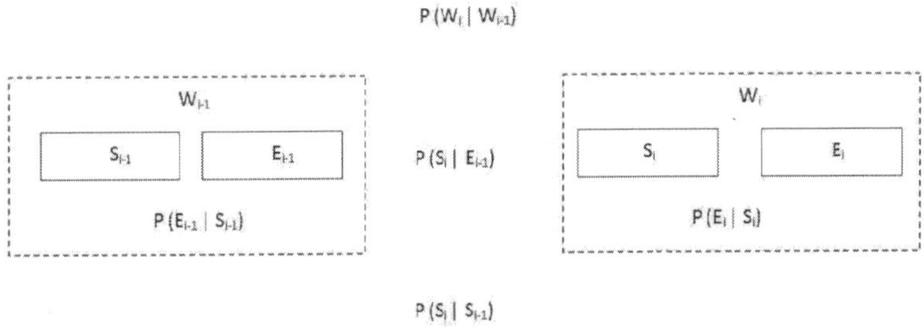

Figure 2. Morpheme based language Modeling

4.2. Syllable Modeling

The importance of syllable as a unit in ASR was felt in early researches starting with (Fujimura 1975) where irregularities in phonemes have been discussed and it has been claimed that a syllable will serve as the viable minimal unit of speech in time domain.

In the paper (Greenberg 1998), it is stated that pronunciation variation in Switchboard corpus is more systematic at the level of a syllable. It has been emphasized that the onset and nucleus of a syllable do not show much contextual dependencies while the coda may still be susceptible to some contextual effects with the following syllable. Greenberg (1998) proposed that syllables are frequently realized in their standard or canonical form but in the case of phones, canonical realization is mostly unusual.

In the paper by Ganapathiraju et al (2001), the first successful robust LVCSR system that used syllable level acoustic unit in telephone bandwidth spontaneous speech is reported. The paper begins with a conjecture that syllable based system would perform better than existing triphone systems and concludes with experimental verification after comparing a syllable based system performance with that of a word-internal and a cross-word triphone system on publicly available databases, viz. Switchboard and Alphadigits. A number of syllable based experiments involving syllables and CI phones, syllables and CD phones, syllables, mono-syllabic words and CD phones have been reported in that paper. However, this system is deficient especially in the integration of syllable and phone models as mixed-word entry. It is because mixing models of different lengths and context might result only in marginal improvements.

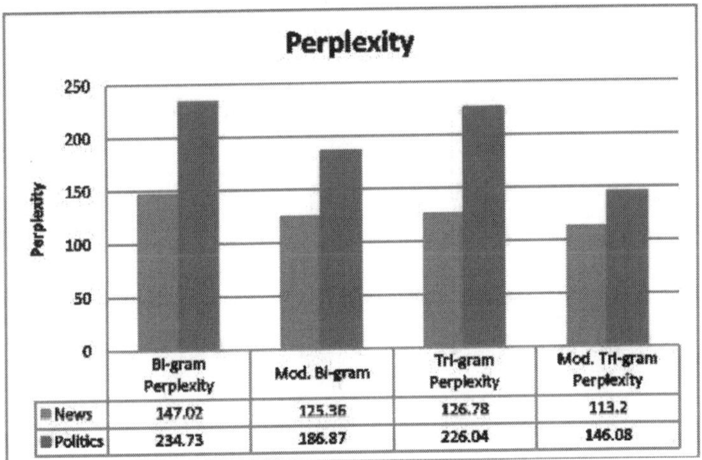

(a) Comparison of Perplexity

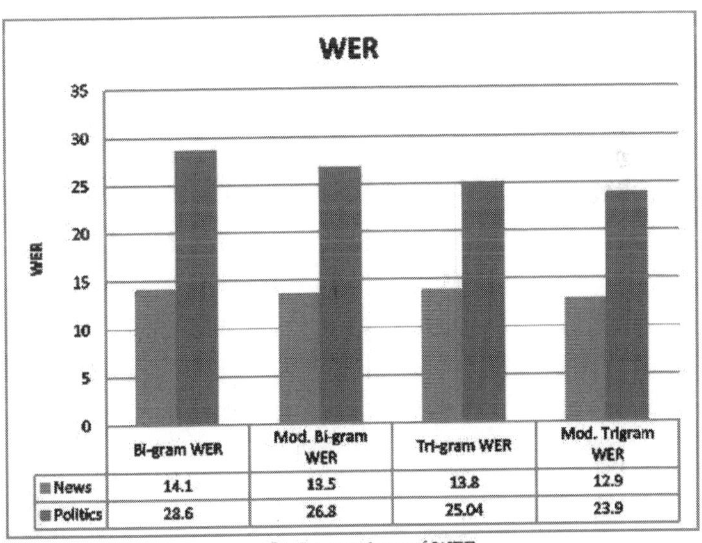

(b) Comparison of WER

Figure 3.

4.2.1. Justification For Using Prosodic Syllable As A Speech Unit

Thangarajan et al (2008a) have proposed a syllable based language model for combating the agglutinative nature of Tamil language. The basic syllable consonant-vowel phono-tactics in Tamil is characterized by a regular expression shown in equation (3). There are constraints on which consonants can appear in each of the three consonant positions and in combination with vowels. With no constraints, the maximum number of syllables will be 183 × 12 = 69,984. However, because of constraints the actual number of possible syllables is in order of magnitude smaller. The number of lexically attested syllable is smaller still. In addition, there are constraints on stress patterning in Tamil words.

$$RE(S) = [C]V[C[C]] \tag{3}$$

Properties that constitute prosody are fundamental frequency or formant f0 (perceived pitch), duration, intensity (perceived loudness) and to some extent vowel quality. Prosodic properties of speech are also used in detection of word boundaries and in other higher tasks of speech understanding like encoding or decoding pragmatic differences like a statement vs. a question, emotion and so on. At the word level, prosodic properties encode lexical tone, lexical stress and lexical pitch accent.

There are two types of prosodic syllables namely Ner-acai and Nirai-acai. Ner-acai is monosyllabic. It may consist of either one short vowel or one long vowel, either of which may be open or closed, i.e. ending in a vowel or consonant(s) respectively. Nirai-acai is always disyllabic with an obligatorily short vowel at first position, while the second phoneme is unrestricted. Like Ner-acai, Nirai-acai may also be of open or closed type. The prosodic syllable representation can take any of the following eight patterns as shown in Table 3. An uninflected Tamil word may comprise one to four prosodic syllables.

4.2.2. Formal Representations Of Prosodic Syllables

Prosodic syllables are composed of phonemes. Based on the linguistic rules tabulated in Table 3, a regular expression can be formulated as shown in equation (4).

$$RE(S) = [SV](SV|LV)[C[C]] \tag{4}$$

This expression describes all the possible patterns of prosodic syllables. In other words, an optional short vowel is followed obligatorily by either a short vowel or a long vowel, and zero or one or two consonants.

Table 3. The Linguistic Rules of Tamil Prosodic Syllables

Description	Pattern	Example (with Romanized Tamil and meaning)
Short vowel, long vowel followed by consonant(s)* (Nirai)	SV + LV + C(s)	புலால் (pulal) (meat)
Short vowel followed by a long vowel, (Nirai)	SV + LV	விழா (vizha) (function)
Two short vowels followed by consonant(s)*, (Nirai)	SV + SV + C(s)	களம் (kaLam) (field)
Two short vowels, (Nirai)	SV + SV	கல (kala) (echo sound)
Short vowel followed by consonant(s)*, (Ner)	SV + C(s)	கல் (kal) (stone)
Long vowel followed by consonant(s)*, (Ner)	LV + C(s)	வாள் (vaL) (sword)
Long vowel, (Ner)	LV	வா (va) (come)
Short vowel, (Ner)	SV	க (ka)

Theoretically, the number of prosodic syllables will be quite larger (of the order of 3,674,160), since there are 90 (18 times 5) short vowels, 126 (18 times 7) long vowels and 18 consonants. But, the actual number will be smaller due to constraints like phono-tactics and morpho-tactics. Hence, it is essential to estimate the number of prosodic syllables with the help of a corpus.

4.2.3. Analysis Of Tamil Text Corpus

In this section, a Tamil text corpus provided by Central Institute for Indian Languages (CIIL) with 2.6 million words is taken and useful statistics about prosodic syllables is collected. This corpus is a collection of Tamil text documents collected from various domains, viz. agriculture, biographies, cooking tips and news articles. A simple algorithm to segment prosodic syllables from a word is proposed whose pseudo code is given below.

```
Function Syllabify (Word[0..n-1])

    k ← 0 // Index of current letter in the WORD
    m ← 0 // Index of syllable array
    // for each letter in the 'Word' categorize it as
    // short vowel, long vowel or consonant
    for k ← 0 to n-1
        if (Word[k] is a short vowel)
                CharCategory[k] ← 0
        else if (Word[k] is a long vowel)
                CharCategory[k] ← 1
        else
                CharCategory[k] ← 2   // it is a consonant
    end for

    for k ← 0 to n-1
        if ((k+2) <= n and CharCategory[k] = 0 and
                 CharCategory[k + 1] = 1 or

                 CharCategory[k + 1] = 0))
                copy(Syllable[m], Word[k], Word[k+1]);
                k ← k + 2;

      else if (CharCategory[k] = 1 || CharCategory[k] = 0)
              copy(Syllable[m], Word[k]);
              k ← k + 1;
      end if

      while(k < n && CharCategory[k] = 2)
         copy(Syllable[m], Word[k]);
              k ← k + 1;
      end while
      m ← m + 1;
   end for
     return m; // returns the no. of syllables; syllables
               // are stored Syllable[]
end function
```

The algorithm works in two stages. Initially, grapheme to phoneme conversion (phonetisation) is done by scanning all the letters of a word and categorizing them as vowels and consonants. The next step of the algorithm combines the letters into syllables with the help of linguistic rules which are presented inTable 3. This step is called syllabification. Syllable patterns are checked from the biggest syllable to the smallest one. The algorithm stores the syllables in an array and returns their count.

After applying the algorithm to the text corpus, the frequency counts of various prosodic syllable patterns were gathered. The algorithm segmented 26,153 numbers of unique prosodic syllables in the corpus. Since the text corpus used here was not clean, it contained a lot of abbreviations, digits and other foreign characters. Therefore, the prosodic syllable patterns with frequency less than 10 were eliminated. Then, it was found that there were only 10,015 numbers of unique prosodic syllables as shown in Table 4.

Table 4. Prosodic Syllables in CIIL Corpus

Details	Frequency
Documents	686
Sentences	455,504
Words	2,652,370
No. of unique prosodic syllables segmented by the algorithm	26,153
No. of unique prosodic syllables validated by the DFA	10,015

4.2.4. Creating Context Independent Syllable Models

A lexicon based on prosodic syllables was created with the aid of the algorithm where every word in the dictionary was segmented into its constituent prosodic syllables. Along with the dictionary, a list of prosodic syllable models and continuous speech with sentence aligned transcription were given as input to the training program. The transcription were force-aligned with Baum-Welch training followed by Viterbi alignment.

In order to keep the complexity low, it was preferable to model CI syllable units with single Gaussian continuous density HMM. The continuous speech was transformed into a sequence of feature vectors. This sequence was matched with the optimal/best concatenated HMM sequence found using Viterbi algorithm. The time stamps of segmented syllable boundaries were obtained as a by-product of Viterbi decoding. The duration of the prosodic syllables was found to vary from 290 *ms* to 315 *ms*. Even though a prosodic syllable is either monosyllabic or disyllabic, the duration was more or less equal to 300 ms on average. This may be due to vowel duration reduction which occurs in non-initial syllables as reported by Asher and Keane (2005).

Based on these considerations, eight states per HMM were decided to be adequate for the experiment.Figure 4 shows the schematic block diagram of a syllable based recognizer.

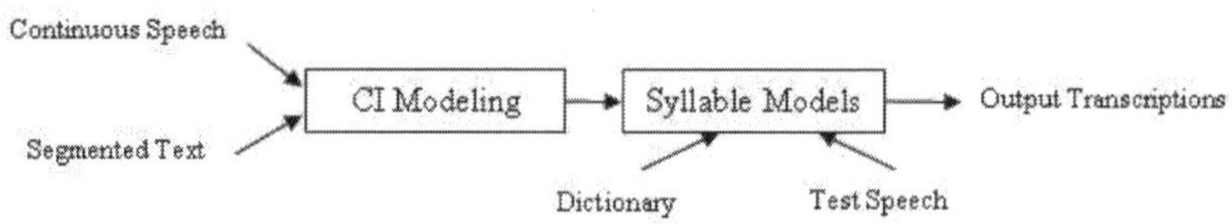

Figure 4. The Syllable Modeling and Recognition System

4.2.5. Results and Discussions

For simplicity, an acoustic model was trained with 1,398 unique prosodic syllables drawn from agriculture domain. These prosodic syllables almost covered the agriculture data and the test set completely. In the experiment, the number of models to be trained was significantly reduced compared to the triphone models. The baseline triphone model had 3,171 numbers of unique triphones extracted from the transcript. The experiment was carried out using syllable based continuous speech recognition for Tamil. The dictionary and the transcripts were segmented into prosodic syllables with the proposed algorithm and models were trained. The syllable based acoustic model was deployed on a conventional continuous speech recognizer and tested with the same test set comprising 400 sentences. When comparing the WER, it is found that the WER of syllable models were considerably reduced (by 10%) compared to word models. However the triphone models performed well with a WER of 9.44%

It was also observed that in the prosodic syllable models, there were larger number of substitution errors than that of insertions and deletions whereas in the case of word models, there was a majority of deletion errors. This comparison is shown in Figure 5. The majority of deletion errors in word models signify OOV rate due to morphological inflections. The OOV words in syllable models significantly got reduced. This proves the fact that syllables are effective as sub-word units according toThangarajan et al (2008b)

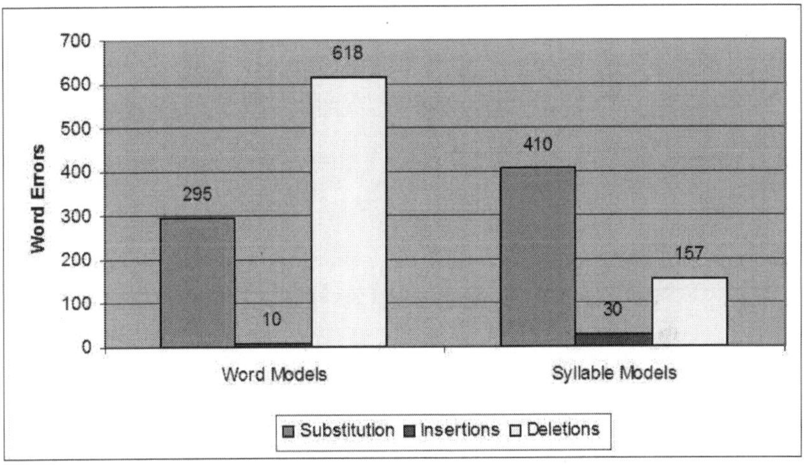

Figure 5. The Types of Word Errors in Word Models and Syllable Models

The increase in WER by 10% approximately in syllable models compared to triphone models can be attributed to the large number of syllables to be modeled with the available limited training set. This also indicates the presence of a little contextual effect between syllables. This is an avenue for future research.

5. SUMMARY

In this chapter, the nature of agglutinative languages is discussed with Tamil language taken as a case study. The inflectional morphology of Tamil language is described in great detail. The challenges that are faced in ASR systems for such languages are highlighted. Two different approaches – enhanced morpheme based languages model and syllable based models - used in ASR for agglutinative languages are elaborated along with their results. The merits and scope for further research is also discussed.

REFERENCES

1. P. Anandan, K. Saravanan, R. Parthasarathy, T. V. Geetha, 2002 'Morphological Analyzer for Tamil', in the proceedings of ICON 2002, Chennai.

2. A. H. Arden, 1934 'A progressive grammar of common Tamil' 4th edition, Christian Literature Society, Madras, India, 59

3. S. Arokianathan, 1981 'Tamil clitics', Dravidian Linguistics Association, Trivandrum, India, 5

4. Asher R.E. and Keane E.L. 2005 'Diphthongs in colloquial Tamil', (Hardcastle W.J. and Mackenzie Beck J. eds.), 141171 .

5. T. Balasubramanian, 1980 'Timing in Tamil', Journal of Phonetics, 8 449467 .

6. O. Fujimura, 1975 'Syllable as a unit of Speech Recognition', IEEE Transactions on Acoustics, Speech and Signal Processing, ASSP-23 , 1 8287 .

7. A. Ganapathiraju, Hamaker. Jonathan, Picone. Joseph, Ordowski. Mark, R. George, Doddington 2001 'Syllable Based Large Vocabulary Continuous Speech Recognition', IEEE Transactions on Speech and Audio Processing, 9 4 358366 .

8. S. Greenberg, 1998 'Speaking in Short Hand- A Syllable Centric Perspective for Understanding Pronunciation Variation', Proceedings of the ESCA Workshop on Modeling Pronunciation Variation for Automatic Speech Recognition, Kekrade, 4756 .

9. F. Harold, Schiffman, 2006 'A Reference Grammar of Spoken Tamil', Cambridge University Press (100521027527

10. Kumar C.S. and Foo Say Wei 2003 'A Bilingual Speech Recognition System for English and Tamil', Proceedings of Joint Conference of the Fourth International Conference on Information, Communications and Signal Processing, 2003 and the Fourth Pacific Rim Conference on Multimedia, 3 16411644 .

11. M. Kumar, N. Rajput, A. Verma, 2004 'A large-vocabulary continuous speech recognition system for Hindi', IBM Journal of Research and Development, 48 5 6, 703715 .

12. Peter. Ladefoged, 1993 'A course in phonetics.' 3rd edition, Fort Worth, TX: Harcourt, Brace, and Jovanovich

13. C. R. Marthandan, 1983 'Phonetics of casual Tamil', Ph.D. Thesis, University of London.

14. S. Nakagawa, Y. Hashimoto, 1988 'A method for continuous speech segmentation using HMM', presented at IEEE International Conference on Pattern Recognition.

15. S. Nakagawa, K. Hanai, K. Yamamoto, N. Minematsu, 1999 'Comparison of syllable-based HMMs and triphone-based HMMs in Japanese speech recognition', Proceedings of International Workshop Automatic Speech Recognition and Understanding, 393396 .

16. Khan. A. Nayeemulla, B. Yegnanarayana, 2001 'Development of Speech Recognition System for Tamil for Small Restricted Task', Proceedings of National Conference on Communication, India.

17. M. Plauche, N. Udhyakummar, C. Wooters, J. Pal, D. Ramachadran, 2006 'Speech Recognition for Illiterate Access to Information and Technology', Proceedings of First International Conference on ICT and Development.

18. S. Rajendran, S. Viswanathan, Kumar. Ramesh, 2003 'Computational Morphology of Tamil Verbal Complex', Language in India, 3 4

19. Rajendran 2004 (2004) 'Strategies in the Formation of Compound Nouns in Tamil', Languages in India, 4 6

20. Rajendran 2006 (2006) 'Parsing In Tamil: Present State of Art', Language in India, 6 8.

21. S. Saraswathi, T. V. Geetha, 2004 'Implementation of Tamil Speech Recognition System Using Neural Networks', Lecture Notes in Computer Science, 3285

22. S. Saraswathi, T. V. Geetha, 2007 'Comparison of Performance of Enhanced Morpheme-based language Model with Different Word-

based Language Models for Improving the Performance of Tamil Speech Recognition System', ACM Transaction on Asian Language Information Processing 6 3 Article 9.

23. F. Soundaraj, 2000 'Accent in Tamil: Speech Research for Speech Technology', In: (Nagamma Reddy K. ed.), Speech Technology: Issues and implications in Indian languages International School of Dravidian Linguistics, Thiruvananthapuram, 246256 .

24. R. Thangarajan, A. M. Natarajan, M. Selvam, 2008a 'Word and Triphone based Approaches in Continuous Speech Recognition for Tamil Language', WSEAS Transactions on Signal Processing, 3 4 2008, 7685 .

25. R. Thangarajan, A. M. Natarajan, (2008b), 'Syllable Based Continuous Speech Recognition for Tamil Language', South Asian Language Review (SALR), Vol. XVIII, No. 1, pp. 71-85.

26. Xuedong Huang, Alex Acero and Hsiao-Wuen Hon 2001 'Spoken Language Processing- A Guide to Theory, Algorithm and System Development', Prentice Hall PTR (0-13022-616-5

Chapter 10

Development of Application Specific Continuous Speech Recognition System in Hindi

Gaurav, Devanesamoni Shakina Deiv, Gopal Krishna Sharma, Mahua Bhattacharya*

Indian Institute of Information Technology & Management, Gwalior, India.

ABSTRACT

Application specific voice interfaces in local languages will go a long way in reaching the benefits of technology to rural India. A continuous speech recognition system in Hindi tailored to aid teaching Geometry in Primary schools is the goal of the work. This paper presents the preliminary work done towards that end. We have used the Mel Frequency Cepstral Coefficients as speech feature parameters and Hidden Markov Modeling to model the acoustic features. Hidden Markov Modeling Tool Kit −3.4 was used both for feature extraction and model generation. The Julius recognizer which is language independent was used for decoding. A speaker independent system is implemented and results are presented.

KEYWORDS

Automatic Speech Recognition; Mel Frequency Cepstral Coefficients; Hidden Markov Modeling

1. INTRODUCTION

To make Information Technology (IT) relevant to rural India, voice access to a variety of computer based services is imperative. Although many speech interfaces are already available, the need is for speech interfaces in local Indian languages. Application specific Hindi speech recognition systems are required to make computer aided teaching, a reality in rural schools. This paper presents the preliminary work done to demonstrate the relevance of a Hindi Continuous Speech Recognition System in primary education.

Automatic speech recognition has progressed tremendously in the last two decades. There are several commercial Automatic Speech Recognition (ASR) systems developed, the most popular among them are Dragon Naturally Speaking, IBM Via voice and Microsoft SAPI. Efforts are on to develop speech recognition systems in different Indian Languages. An isolated word Hindi ASR for small vocabulary is developed and evaluated in [1]. An effort to increase the recognition accuracy of Hindi ASR by online speaker adaptation has been reported in [2]. It is demonstrated that Maximum Likelihood Linear Regression (MLLR) transform based adaptation transforms the acoustic models in such a way that the difference between test and training conditions is reduced, resulting in better performance.

A general approach to identifying feature vectors that effectively distinguish gender of a speaker from Hindi vowel phoneme utterances has been presented in [3,4]. Centre for Development of Advanced computing has developed a domain specific speaker independent continuous speech recognition system for Hindi using Julius recognition engine [5]. They also have built a Hindi ASR for travel domain [6] giving encouraging recognition accuracy.

State likelihood evaluation in Hidden Markov model (HMM) using mixture of Gaussians is one problem that needs to be solved. A novel method using Gaussian Mixture Model (GMM) for statistical pattern classification is suggested to reduce computational load [7]. Development of speech interfaces in Hindi for IT based services is a work in progress [8]. Efforts to compensate for different accents in Hindi are also explored in [9]. Apart from Hindi ASR, speech recognition systems are being developed in other languages like Arabic, Malayalam, Tamil, Bengali, Telugu, etc. [10-14].

IBM Research Laboratory of India has developed a Hindi Speech Recognition system which has been trained on 40 hours of audio data and has a trigram language model that is trained with 3 million words [15]. Efforts are on to develop large speech databases in various Indian Languages for Large Vocabulary Speech Recognition Systems [16]. SRI Language Model (SRILM) extensible toolkit is discussed in [17] which can be used for developing Language model. This toolkit has been used in developing language model for large vocabulary systems in Hindi.

Hidden Markov Model provides an elegant statistical framework for modeling speech patterns and is the most widely used technique [18,19]. Recently the hybrid HMM and Artificial Neural Network (ANN) framework is also used in an effort to overcome the challenges posed by speech variability due to physiological differences, style variability due to co-articulation effects, varying accents, emotional states, context variability etc [20].

Another method to handle the problem of changes in the acoustic environment or speaker specific voice characteristics is by adapting the statistical models of a speech recognizer and speaker tracking. Combining speaker adaptation and speaker tracking may be advantageous, because it allows a system to adapt to more than one user at the same time. Authors in [21] have extended a standard speech recognizer by combining speaker specific speech decoding with speaker identification in an efficient manner. Approximately 20% relative error rate reduction and about 94.6% identification rate are reported.

The system presented here is an application specific Continuous Speech Recognizer in Hindi. It is restricted to the task of computer-aided teaching of Geometry at primary school level. The paper is organized as follows. Section 2 describes the architecture of the speech recognition system with the function of each module. Section 3 explains the training methodology of developing the proposed Hindi CSR. Section 4 details the testing of the system. The results are discussed in Section 5. Section 6 concludes with future direction of the work.

2. AUTOMATIC SPEECH RECOGNITION SYSTEM

Speech recognition is the process of converting an acoustic signal, captured by a microphone or a telephone into a set of words. The recognized words can be the final result for applications such as commands and control, data entry, and document preparation. They can also serve as the input to further linguistic processing in order to achieve speech understanding. **Figure 1** shows the block diagram of a state of the art automatic speech recognition system.

Speech signal is analog. In the first place analog electrical signals are converted to digital signals. This is done in two steps, sampling and quantization. So a typical representation of a speech signal is a stream of 8-bit numbers at the rate of 10,000 numbers per second. Once the signal conversion is complete, background noise is filtered to keep signal to noise ratio high. The signal is pre-emphasized and then speech parameters are extracted.

2.1. Feature Extraction

Mel-Frequency Cepstral Coefficients (MFCCs) are widely used features for automatic speech recognition systems to transform the speech waveform into a sequence of discrete acoustic vectors.

The MFCC technique makes use of two types of filter, namely, linearly spaced filters and logarithmically spaced filters. The Mel frequency scale has linear frequency spacing below 1000 Hz and a logarithmic spacing above 1000 Hz. In the sound processing, the Mel-frequency cepstrum is a representation of the short-term power spectrum of a sound, based on a linear cosine transform of a log power spectrum on a nonlinear Mel-frequency scale.

The procedure by which the Mel-frequency cepstral coefficients are obtained consists of the following steps. **Figure 2** depicts the procedure of extracting MFCC feature vectors from speech.

The signal is passed through a filter which emphasizes higher frequencies. This process will increase the energy of the signal at higher frequency.

The Pre-emphasis of the speech signal is realized with this simple FIR filter

$$H(z) = 1 - az^{-1} \quad (1)$$

where a is from interval [0.9, 1].

The digitized speech is segmented into frames with a length within the range of 10 to 40 ms. The segment of waveform used to determine each parameter vector is usually referred to as a window.

The Hamming window which is used for the purpose is defined by the equation

$$w(n) 0.54 - 0.46 \cos\left[2\pi n/(N-1)\right] \quad (2)$$

where, $0 \le n \le N-1$

N = number of samples in each frame.

Let Y(n) = Output signal and X(n) = input signal The result of windowing the signal is

$$Y(n) = X(n) W(n) \quad (3)$$

Next, the Fast Fourier transform (FFT) is used to convert each frame of N samples from time domain into frequency domain. Thus the components of the magnitude spectrum of the analyzed signal are calculated.

$$Y(\omega) = \mathrm{FFT}\left[h(t) * x(t)\right] = H(\omega) X(\omega) \quad (4)$$

The most important step in this signal processing is Mel-frequency transformation. Compensation for nonlinear perception of frequency is implemented by the bank of triangular band filters with the linear distribution of frequencies along the so called Mel-frequency range. Linear deployment of filters to Mel-frequency axis results in a non-linear distribution for the standard frequency axis in hertz. Definition of the Mel-frequency range is described by the following equation.

$$f_{mel} = 2595 \log_{10}(1 + f/100)\, Hz \quad (5)$$

where f is frequency in linear range and f_{mel} the corresponding frequency in nonlinear Mel-frequency range.

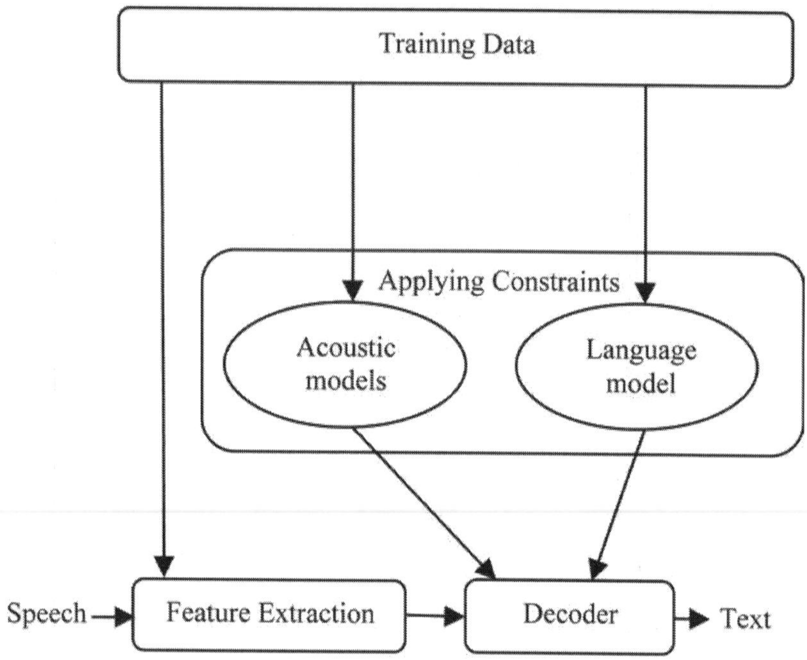

Figure 1. Automatic speech recognition system.

The Mel spectrum coefficients and their logarithm are real numbers. Hence they can be converted to the time domain using the discrete cosine transform (DCT). The result is the Mel Frequency Cepstral Coefficients. The cepstral representation of the speech spectrum provides a good representation of the local spectral properties of the signal for the given frame analysis.

$$c_n = \sqrt{2/K} \sum_{j=1}^{N} \left(\log_{mj}\right) \cos\left(\pi n \left(j - 0.5\right)/K\right) \tag{6}$$

n = number of Mel-frequency cepstral coefficients K = number of Mel-frequency band filters (filter bank channels) in the bank of filters.

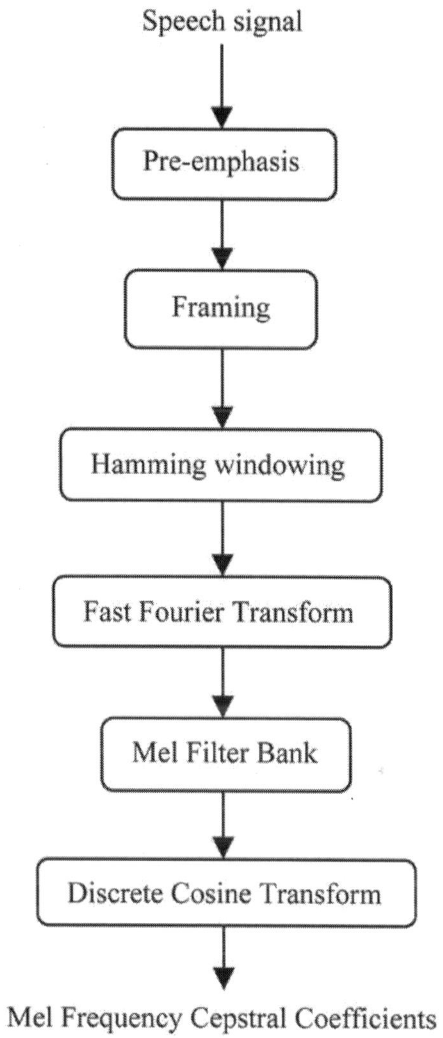

Figure 2. Flow chart of MFCC feature extraction.

2.2. The Acoustic Model

In a statistical framework for speech recognition, the problem is to find the most likely word sequence, which can be described by the equation

$$\hat{W} = \arg_w \max P(W/X) \tag{7}$$

Applying the Bayes' equation, we get

$$\hat{W} = \arg_w \max P(W/X)P(W) \qquad (8)$$

The term P(X/W) in the above equation can be realized by the Acoustic model. An acoustic model is a file that contains a statistical representation of each distinct sound that makes up a spoken word. It contains the sounds for each word found in the Language model.

The speech recognition system implemented here uses Hidden Markov Models (HMM) for representing speech sounds. A HMM is a stochastic model. A HMM consists of a number of states, each of which is associated with a probability density function. The model parameters are the set of probability density functions, and a transition matrix that contains the probability of transitions between states.

HMM-based recognition algorithms are classified into two types, namely, phoneme level model and word-level model. The word-level HMM has excellent performance at isolated word tasks and is capable of representing speech transitions between phonemes. However, each distinct word has to be represented by a separate model which leads to extremely high computation cost (which is proportional to the number of HMM models). The phoneme model on the other hand can help reproduce a word as a sequence of phonemes. Hence new words can be added to the dictionary without necessitating additional models. Hence phoneme model is considered more suitable in applications with large sized vocabularies and where addition of word is a essential possibility.

The phoneme model is used here. The MFCC features extracted from speech and the associated transcriptions are used to estimate the parameters of HMM based acoustic models that represent phonemes. The iterative process of estimating and re-estimating the parameters to achieve a reasonable representation of the speech unit is called ASR system training. The training procedure involves the use of forward-backward algorithm.

2.3. The Language Model

The term P(W) in Equation (2) represents the a priori probability of a word sequence based on syntax, semantics and pragmatics of the language to be recognized. It can be realized by the Language Model which contains a list of words and their probability of occurrence in a given sequence, and is independent of the acoustic signal. The probability of a word sequence is given below.

$$p(w_1, w_2, w_3, w_4, w_5, \cdots, w_n) = p(W) \tag{9}$$

By Chain rule the probability of nth word is:

$$p(w_1^n) = p(w_1)\, p(w_2|w_1)\, p(w_2|w_1^2) \cdots p(w_n|w_1^{n-1}) \tag{10}$$

$$p(w_1^n) = \prod_{k=1}^{n} p(w_k|w_1^{k-1}) \tag{11}$$

Language Model or Grammar essentially defines constraints on what the Speech Recognition Engine can expect as input.

2.4. The Recognizer

Recognizer is a Software program that takes the sounds spoken by a user and searches the Acoustic Model for the equivalent sounds. When a match is made, the Decoder determines the phoneme corresponding to the sound. It keeps track of the matching phonemes until it reaches a pause in the user's speech. It then searches the Language Model or Grammar file for the equivalent series of phonemes. If a match is made it returns the text of the corresponding word or phrase to the calling program.

3. THE TRAINING METHODOLOGY

The two major stages involved in the process of Speech Recognition are the training of the ASR and the Testing. The training phase involves the following steps.

3.1. The Database

The text corpus consists of chosen application specific sentences, pertaining to teaching Geometry to children. Forty three distinct Hindi sentences about shape geometry using 29 distinct Hindi phonemes were designed as the text corpus. These sentences were spoken in a continuous fashion and recorded using good quality microphones under office noise conditions.

The Wave-surfer software was used for recording. The training corpus contains 1806 utterances spoken by 12 females and 18 males. All the speakers are natives of the Hindi heart-land of India, educated and in the age group of 18 to 30.

3.2. Phone Set

Phoneme is the basic unit of sound in any language. Hindi belongs to the Indo Aryan family of languages and is written in the Devanagari script. There are 11 vowels and 35 consonants in standard Hindi. In addition, five Nukta consonants are also adopted from Farsi/Arabic sounds. The phone set that is used here to develop the application specific speech recognition system for Hindi language uses only 29 of the 60 used in large vocabulary systems.

3.3. Lexicon

The pronunciation dictionary (lexicon) contains all the distinct words in the corpus and its corresponding pronunciation given as a string of phonemes. Some sample entries are given in**Table 1**. The pronunciation dictionary is case insensitive. This dictionary includes entries for the beginning-of-sentence and the end-of-sentence tokens and respectively as well as the silence.

3.4. Transcription

The transcription file contains the sentences or utterances of the spoken text and the corresponding audio files in the following format. Each word in the transcription file is present in the pronunciation lexicon.

3.5. Parameterization of Speech Data

The digitized speech signal is subjected to first order preemphasis applied using a coefficient of 0.97. The signal is then segmented into frames and hamming windowed. The HMM Tool Kit (HTK) [22] was used to parameterize the raw speech waveforms into sequences feature vectors.

Mel Frequency Cepstral Coefficients (MFCCs) are derived from FFT-based log spectra. Coding was performed using the tool HCopy configured to automatically convert its input into MFCC vectors. A configuration file specifies all of the conversion parameters [22]. A typical configuration file is seen in **Figure 3**.

The target parameters are to be MFCC using C_0 as the energy component. The standard 39 dimension MFCC, delta and acceleration feature vector is computed for the 16 kHz sampled signals at 10 ms intervals (100 ns). Mel scaled 26 filter banks spanning the 8 kHz frequency range are used for computation of MFCCs.

The output was saved in compressed format, and a crc checksum added. The 39-demensional feature vector consists of 13 Mel Scale Cepstral Coefficients and their first and second derivates. A sample MFCC file is shown below in **Figure 4**.

3.6. Acoustic Model Generation

The speech recognition system implemented here employs Hidden Markov Model (HMM) for representing speech sounds. A HMM consists of a number of states, each of which is associated with a probability density function. The parameters of a HMM comprises of the parameters of the set of probability density functions, and a transition matrix that contains the probability of transition between states.

Table 1. Pronunciation lexicon.

आयत	aa y ax t sp
बैगनी	b ae g n iy sp
बनाओ	b ax n aa ow sp
चतुर्भुज	ch ax t uh r b hh uh jh sp
एक	ey k sp
हरा	hh ax r aa sp
काला	k aa l aasp
लाल	l aa l sp
नारंगी	n aa r ax ng iy sp
नीला	n iy l aa sp
पीला	p iy l aa sp
सफ़ेद	s ax f eh ey dh sp
समान्तर	s ax m aa n t ax r sp
सम्बाहू	s ax m b aa hh uh sp

The MFCC feature vectors extracted from speech signals and their associated transcriptions are used to estimate the parameters of HMMs. This process is called ASR system training. HMM Tool Kit, HTK-3.4 was used for training models over 29 context-dependent Hindi phonemes used in the chosen application. The basic acoustic units are context dependent phonemes, that is, tri-phones modeled by left-to-right, 5-state, HMMs.

The output probability distributions of states were represented by Gaussian mixture densities. For every state of every phoneme 256 global Gaussian density functions were used to generate Gaussian mixtures.

Prototype models are built using the flat start approach. With the exception of the transition probabilities, all of the HMM parameters given in the prototype definition are ignored. The purpose of the prototype definition is only to specify the overall characteristics and topology of the HMM. The actual parameters will be computed later.

```
# Coding parameters
SOURCEFORMAT = WAV
TARGETKIND = MFCC_0_D_A
TARGETRATE = 100000.0
SAVECOMPRESSED = T
SAVEWITHCRC = T
WINDOWSIZE = 250000.0
USEHAMMING = T
PREEMCOEF = 0.97
NUMCHANS = 26
CEPLIFTER = 22
NUMCEPS = 12
ENORMALISE = F
```

Figure 3. A typical configuration file.

A prototype model is shown in **Figure 5**.

These models were further refined by applying nine iterations of the standard Baum-Welch embedded training procedure. These models are then converted to tri-phone models and two iterations of Baum-Welch training procedure are applied, then the states are tied using decision tree based approach and iterations of Baum-Welch training procedure are applied. **Figure 6** shows the training procedure.

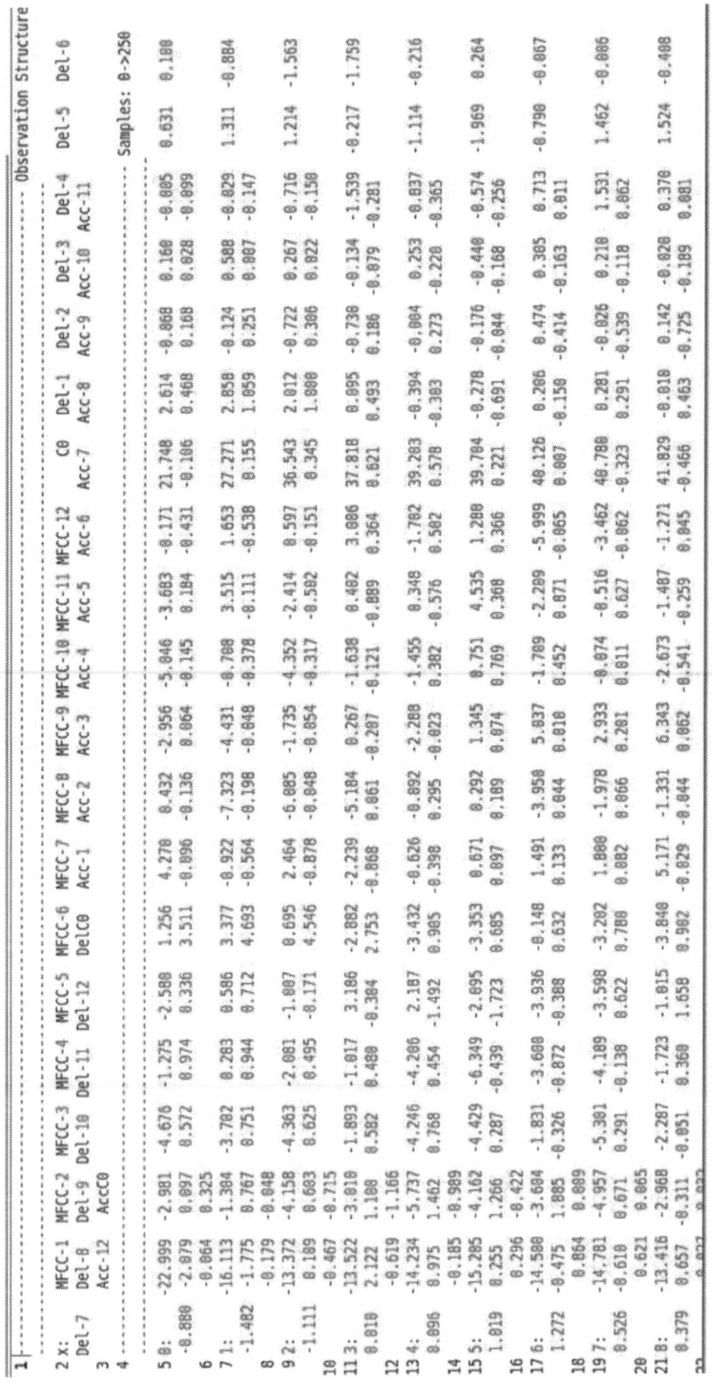

```
1 ------------------------------------------------------------------------------------------------ Observation Structure
  ------------------------------------------------------------------------------------------------
2 x:      MFCC-1  MFCC-2  MFCC-3  MFCC-4  MFCC-5  MFCC-6  MFCC-7  MFCC-8  MFCC-9 MFCC-10 MFCC-11 MFCC-12      C0   Del-1   Del-2   Del-3   Del-4   Del-5   Del-6
   Del-7   Del-8   Del-9  Del-10  Del-11  Del-12   DelC0   Acc-1   Acc-2   Acc-3   Acc-4   Acc-5   Acc-6   Acc-7   Acc-8   Acc-9  Acc-10  Acc-11
3         Acc-12   AccC0
4 ------------------------------------------------------------------------------------------------ Samples: 0->250
  ------------------------------------------------------------------------------------------------
5 0:     -22.999  -2.981  -4.676  -1.275  -2.580   1.256   4.270   0.432  -2.956  -5.046  -3.683  -0.171  21.748   2.614  -0.068   0.160  -0.005   0.631   0.100
  -0.880  -2.079   0.097   0.572   0.974   0.336   3.511  -0.096  -0.136   0.064  -0.145   0.184  -0.431  -0.106   0.468   0.168   0.028  -0.099
6         -0.064   0.325
7 1:     -16.113  -1.304  -3.702   0.283   0.586   3.377  -0.922  -7.323  -4.431  -0.708   3.515   1.653  27.271   2.858  -0.124   0.588  -0.029   1.311  -0.884
  -1.482  -1.775   0.767   0.751   0.944   0.712   4.693  -0.564  -0.198  -0.048  -0.378  -0.111  -0.538   0.155   1.059   0.251   0.007  -0.147
8         -0.179  -0.048
9 2:     -13.372  -4.158  -4.363  -2.081  -1.007   0.695   2.464  -6.085  -1.735  -4.352  -2.414   0.597  36.543   2.012  -0.722   0.267  -0.716   1.214  -1.563
  -1.111   0.189   0.603   0.625   0.495  -0.171   4.546  -0.878  -0.048  -0.054  -0.317  -0.502  -0.151   0.345   1.000   0.306   0.022  -0.150
10        -0.467  -0.715
11 3:    -13.522  -3.010  -1.893  -1.017   3.186  -2.882  -2.239  -5.184   0.267  -1.638   0.402   3.086  37.818   0.095  -0.730  -0.134  -1.539  -0.217  -1.759
   0.010   2.122   1.100   0.582   0.480  -0.304   2.753  -0.868   0.061  -0.207  -0.121  -0.889   0.364   0.621   0.493   0.186  -0.079  -0.281
12        -0.619  -1.166
13 4:    -14.234  -5.737  -4.246  -4.206   2.187  -3.432  -0.626  -0.892  -2.288  -1.455   0.348  -1.702  39.283  -0.394  -0.004   0.253  -0.837  -1.114  -0.216
   0.096   0.975   1.462   0.768   0.454  -1.492   0.905  -0.398   0.295  -0.023   0.382  -0.576   0.502   0.578  -0.303   0.273  -0.220  -0.365
14        -0.185  -0.989
15 5:    -15.205  -4.162  -4.429  -6.349  -2.095  -3.353   0.671   0.292   1.345   0.751   4.535   1.280  39.704  -0.278  -0.176  -0.440  -0.574  -1.969   0.264
   1.019   0.255   1.266   0.287  -0.439  -1.723   0.685   0.097   0.189   0.074   0.769   0.368   0.366   0.221  -0.691  -0.044  -0.168  -0.256
16         0.296  -0.422
17 6:    -14.580  -3.604  -1.831  -3.600  -3.936  -0.148   1.491  -3.950   5.037  -1.709  -2.209  -5.999  40.126   0.206   0.474   0.305   0.713  -0.790  -0.067
   1.272  -0.475   1.885  -0.326  -0.872  -0.388   0.632   0.133   0.044   0.010   0.452   0.871  -0.065   0.007  -0.150  -0.414  -0.163   0.011
18         0.864   0.009
19 7:    -14.781  -4.957  -5.301  -4.109  -3.598  -3.202   1.000  -1.978   2.933  -0.074  -0.516  -3.462  40.780   0.281  -0.026   0.210   1.531   1.462  -0.006
   0.526  -0.610   0.671   0.291  -0.138   0.622   0.780   0.082   0.066   0.201   0.011   0.627  -0.062  -0.323   0.291  -0.539  -0.118   0.062
20         0.621   0.065
21 8:    -13.416  -2.968  -2.287  -1.723  -1.015  -3.840   5.171  -1.331   6.343  -2.673  -1.487  -1.271  41.829  -0.010   0.142  -0.020   0.370   1.524  -0.408
   0.379   0.657  -0.311  -0.051   0.360   1.658   0.902  -0.029  -0.044   0.062  -0.541  -0.259   0.045  -0.466   0.463  -0.725  -0.189   0.081
```

Figure 4. Screenshot of an MFCC file.

4. EVALUATION METHODOLOGY

The performance of the ASR is tested while transcribing unknown utterances. A database which is not used for training the system is called unseen data. The test data here is an exclusive set consisting of 344 unseen utterances spoken by 8 speakers (4 males and 4 females) each speaking 43 sentences.

4.1. Role of Language Model

In speech recognition the Language Model is used for the task of finding word boundaries, that is, segmentation. The language model or grammar which is an a priori knowledge of the syntax, semantics and pragmatics of the language in question, helps decode the sequence of phonemes into different words in a sentence. An example is given in **Figure 7**.

Here the constraints applied by the Language model helps the Recognizer in decoding the phoneme sequence into words. We have generated our own language model.

4.2. The Recognizer

The decoder used for recognition is Julius. Since Julius itself is a language-independent decoding program [23], we can make a recognizer of any language if given an appropriate language model and acoustic model for the target language. The recognition accuracy largely depends on the models. Julius is a real-time, high-speed, accurate recognition engine based on 2-step strategy. It works in two steps. The first step is a high-speed approximate search, which uses a 2-gram frame synchronous beam searching algorithm. In the first step, a treestructured lexicon assigned with the language model probabilities was applied. Pre-computed unigram factoring values are assigned to the intermediate nodes and bi-gram probabilities on the word-end nodes.

```
~o
<STREAMINFO> 1 39
<VECSIZE>
39<NULLD><MFCC_D_A_0><DIAGC>
~h "proto
<BEGINHMM>
<NUMSTATES> 5
<STATE> 2
<MEAN> 39
-7.055892e+00 -2.760827e+00 -1.855420e+00 .......
<VARIANCE> 39
3.614088e+01 4.895053e+01 6.375173e+01 .......
<GCONST> 1.185559e+02
<STATE> 3
<MEAN> 39
-7.055892e+00 -2.760827e+00 -1.855420e+00 ....
<VARIANCE> 39
3.614088e+01 4.895053e+01 6.375173e+01 .......
<GCONST> 1.185559e+02
<STATE> 4
<MEAN> 39
-7.055892e+00 -2.760827e+00 -1.855420e+00 ......
<VARIANCE> 39
3.614088e+01 4.895053e+01 6.375173e+01 .......
<GCONST> 1.185559e+02
<TRANSP> 5
0.000000e+00 1.000000e+00 0.000000e+00
0.000000e+00 0.000000e+00
0.000000e+00 6.000000e-01 4.000000e-01
0.000000e+00 0.000000e+00
0.000000e+00 0.000000e+00 6.000000e-01
4.000000e-01 0.000000e+00
0.000000e+00 0.000000e+000.000000e+00
7.000000e-01 3.000000e-01
0.000000e+00 0.000000e+00 0.000000e+00
0.000000e+00 0.000000e+00
<ENDHMM>
```

Figure 5. A prototype model.

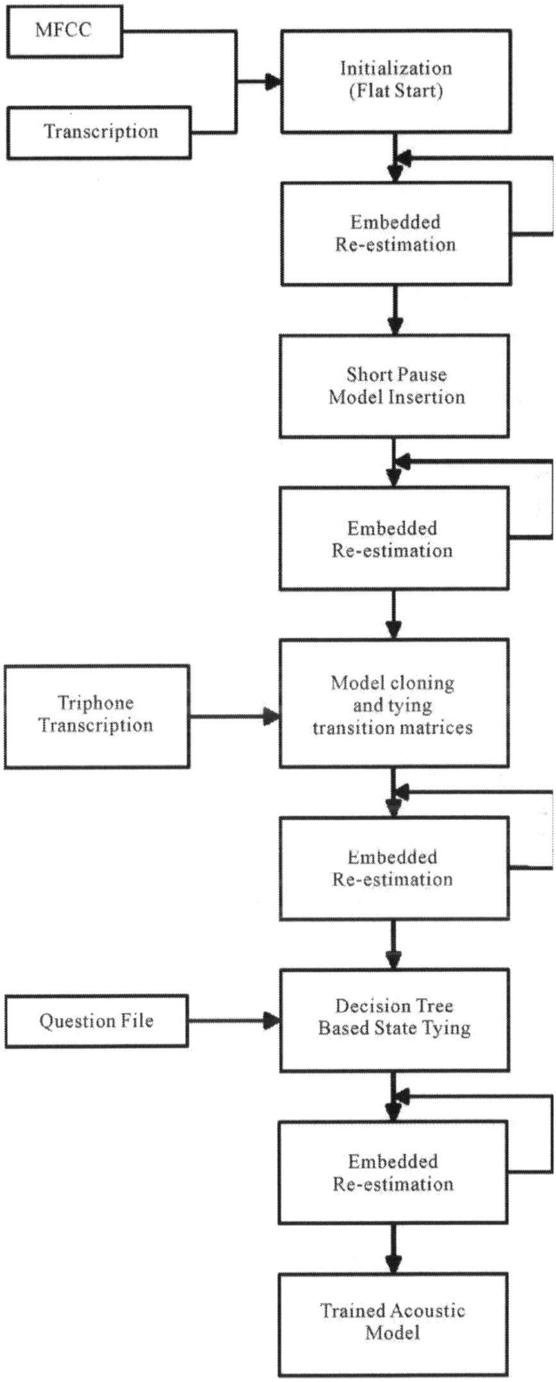

Figure 6. Acoustic model training methodology.

The second step is a high precision trigram N-best stack decoding. The tree trellis search in the second pass recovers the degradation caused by the rough approximation in the first step. Julius adopts acoustic models in HTK ASCII format, pronunciation dictionary in almost HTK format, and word 3-gram language models in ARPA standard format (forward 2-gram and reverse 3- gram trained from same corpus). The following is a sample output for one of our test utterances.

4.3. The Evaluation Parameters

Finally, the recognition accuracy of the Speaker Independent ASR system and the percentage of correct words and percentage of correct sentences were calculated using the following formulae.

$$\%\text{correct} = H/N \tag{12}$$

where, H = Number of labels (sentences here) correctly recognized N = Total number of labels

$$\%\text{Recognition Accuracy} = (N - D - S - I)/N \tag{13}$$

D = Number of unrecognized/missed words. (Deletion errors)

S = Number of times a word was misrecognized as another word (Substitution errors)

I = Number of extra words inserted between correctly recognized words (Insertion errors)

N = Total number of words or sentences

5. RESULTS AND DISCUSSION

The system was trained with 1806 Hindi utterances (sentences) spoken by 18 males and 12 females. The performance of the system was evaluated both for seen and unseen speech data. All the 43 distinct sentences for which the system was trained were uttered by 8 persons (4 males and 4 females). A total of 316 test (unseen) utterances and 1371 seen utterances were used in testing. The % of correct sentences and Recognition Accuracy were calculated using formulae given above and results are shown in **Table 2**.

[ey k sp s ax m aa n t ax r sp ch ax t uh r b hh uh jh sp b ax n aa ow sp]

एक समान्तर चतुर्भुज बनाओ

Figure 7. The Recognition of series of phonemes into series of words.

The Recognition Accuracy for males is better as expected as the ASR is speaker independent and the male speech data is more than that of females. The amount of training data must be increased to achieve better speaker independent model.

6. CONCLUSION AND FUTURE WORK

We have proposed an approach to implement a continuous speech recognition system in Hindi customized for computer aided teaching of geometry. We have used the MFCC as speech feature parameters and HMM to model the acoustic features. HTK-3.4 was used both for feature extraction and model generation. The Julius recognizer which is language independent was used for decoding.

The present work was limited to 29 phonemes of Hindi. It is mostly demonstrative in nature. The future endeavor will be to make the system full-fledged by increasing vocabulary to include all required words and all the Hindi phonemes. A phonetically balanced and rich database for the said application will be created and used.

More training data will be collected and used to improve the speaker Independent system. Methods to improve the Recognition rate of the speaker Independent system will be studied and experimented. Feature sets other than MFCC will be tested for reducing speaker and other variability

Table 2. Recognition accuracies and % of correct words for speakers in training and test sets.

Speakers		% Correct Sentences	% Recognition Accuracy (Words)
	Male (18)	76.84	92.72
Training (Seen)	Female (12)	60.28	84.9
	All (30)	68.56	88.81
Test (Unseen)	All (8 = 4M + 4F)	42.72	79.11

REFERENCES

1. K. Kumar and R. K. Agarwal, "Hindi Speech Recognition System Using HTK," International Journal of Computing and Business Research, Vol. 2, No. 2, 2011, ISSN (Online): 2229-6166.

2. G. Sivaraman and K. Samudravijaya, "Hindi Speech Recognition and Online Speaker Adaptation," Proceedings of ICTSM 2011, Vol. 145, 2011, pp. 233-238.

3. D. ShakinaDeiv, Gaurav and M. Bhattacharya, "Automatic Gender Identification for Hindi Speech Recognition," International Journal of Computer Applications, Vol. 31, No. 5, 2011, pp. 1-8.

4. R. K. Aggarwal and M. Dave, "Implementing a Speech Recognition System Interface for Indian Language," Proceedings of the IJCNLP-2008 Workshop on NLP for Less Privileged Languages, Hyderabad, January 2008, pp. 105-112.

5. R. Mathur, Babita and A. Kansal, "Domain Specific Speaker Independent Continuous Speech Recognition Using Julius," ASCNT 2010.

6. S. Arora, B. Saxena, K. Arora and S. S. Agarwal, "Hindi ASR for Travel Domain," Oriental COCOSDA 2010 Proceedings Centre for Development of Advanced Computing, Noida, 24-25 November 2010.

7. R. K. Aggarwal and M. Dave, "Fitness Evaluation of Gaussian Mixtures in Hindi Speech Recognition System," 2010 First International Conference on Integrated Intelligent Computing, Bangalore, 5-7 August 2010, pp. 177- 183.

8. K. Samudravijaya, "Hindi Speech Recognition," Journal Acoustic Society of India, Vol. 29, No. 1, 2009, pp. 385- 393.

9. K. Malhotra and A. Khosla, "Automatic Identification of Gender & Accent in Spoken Hindi Utterances with Regional Indian Accents," IEEE Spoken Language Technology Workshop, Goa, 15-19 December 2008, pp. 309- 312.

10. R. Gupta, "Speech Recognition for Hindi," M. Tech. Project Report, Department of Computer Science and Engineering, Indian Institute of Technology, Bombay, Mumbai, 2006.

11. B. A. Q. Al-Qatab and R. N. Ainon, "Arabic Speech Recognition Using Hidden Markov Model Toolkit (HTK)," International Symposium in Information Technology, Kuala Lumpur, 15-17 June 2011, pp. 557-562.

12. C. Kurian and K. Balakrishnan, "Speech Recognition of Malayalam Numbers," World Congress on Nature & Biologically Inspired Computing, Coimbatore, 9-11 December 2009, pp. 1475-1479.

13. R. Syama and S. M. Idikkula, "HMM Based Speech Recognition System for Malayalam," The International Conference on Artificial Intelligence, 2008Monte Carlo Resort, Las Vegas, 14-17 July 2008.

14. P. G. Deivapalan and H. A. Murthy, "A Syllable-Based Isolated Word Recognizer for Tamil Handling OOV Words," The National Conference on Communications, Indian Institute of Technology Bombay, 1-3 February 2008, pp. 267-271.

15. C. Neti, N. Rajput and A. Verma, "A Large Vocabulary Continuous Speech Recognition System for Hind," IBM Research and Development Journal, September 2004.

16. G. Anumanchipalli, R. Chitturi, S. Joshi, R. Kumar, S. P. Singh, R. N. V. Sitaram and S. P. Kishore, "Development of Indian Language Speech Databases for Large Vocabulary Speech Recognition Systems," Proceedings of International Conference on Speech and Computer (SPECOM), Patras, October 2005.

17. A. Stolcke, "SRILM—An Extensible Language Modeling Toolkit," Proceedings of the 7th International Conference on Spoken Language Processing, 2002, pp. 901-904. http://www.speech.sri.com/

18. L. R. Rabiner, "A Tutorial on Hidden Markov Models and Selected Applications in Speech Recognition," Proceedings of the IEEE, Vol. 77, No. 2, 1989, pp. 257-286.

19. C. J. Legetter, "Improved Acoustic Modeling for HMMs Using Linear Transformations," Ph.D. Thesis, University of Cambridge, Cambridge, 1995.

20. M. Benzeghiba, R. De Mori, O. Deroo, S. Dupont, T. Erbes, D. Jouvet, L. Fissore, P. Laface, A. Mertins, C. Ris, R. Rose, V. Tyagi and C. Wellekens, "Automatic Speech Recognition and Speech Variability: A Review," Speech Communication, Vol. 49, No. 10-11, 2007, pp. 763-786.

21. T. Herbig, F. Gerl, W. Minker and R. Haeb-Umbach, "Adaptive Systems for Unsupervised Speaker Tracking and Speech Recognition," Evolving Systems, Vol. 2, No. 3, 2011, pp. 199-214.

22. Steve Young, et al., "The HTK Book," http://htk.eng.cam.ac.uk/docs/docs.shtml

23. A. Lee, T. Kawahara and K. Shikano, "Julius—An Open Source Real-Time Large Vocabulary Recognition Engine," Proceedings of 7th European Conference on Speech Communication and Technology, 2001.

Index